"十二五"普通高等教育本科国家级规划教材

中国高等学校计算机科学与技术专业（应用型）规划教材

C++ STL基础及应用
（第2版）

刘德山 金百东 编著

清华大学出版社
北京

内容简介

本书全面而又系统地介绍标准模板库(STL)泛型应用开发技术,基础知识部分包括模板、迭代器、输入输出流、字符串、函数对象、通用容器、非变异算法、变异算法、排序等,集成应用部分包括STL算法的综合应用、在数据结构中的应用、在Visual C++上的应用等。本书从应用出发,每章都包含大量示例和详细的结果分析,旨在使读者学会STL各个知识体系的应用方法,体会STL思维的巧妙之处。对某些稍难示例的设计思想也做了详细说明。

本书可作为专业技术人员、大专院校计算机专业的本科生、研究生学习C++泛型编程的教材或参考书,对编写Java泛型程序也有一定的指导意义。

本书封面贴有清华大学出版社防伪标签,无标签者不得销售。
版权所有,侵权必究。举报: 010-62782989,beiqinquan@tup.tsinghua.edu.cn。

图书在版编目(CIP)数据

C++ STL基础及应用/刘德山,金百东编著. —2版. —北京:清华大学出版社,2015(2024.10重印)
中国高等学校计算机科学与技术专业(应用型)规划教材
ISBN 978-7-302-40035-6

Ⅰ. ①C… Ⅱ. ①刘… ②金… Ⅲ. ①C语言—程序设计—高等学校—教材 Ⅳ. ①TP312

中国版本图书馆CIP数据核字(2015)第086760号

责任编辑:谢 琛
封面设计:常雪影
责任校对:白 蕾
责任印制:丛怀宇

出版发行:清华大学出版社
网　　址:https://www.tup.com.cn,https://www.wqxuetang.com
地　　址:北京清华大学学研大厦A座　　　邮　　编:100084
社　总　机:010-83470000　　　　　　　　邮　　购:010-62786544
投稿与读者服务:010-62776969,c-service@tup.tsinghua.edu.cn
质量反馈:010-62772015,zhiliang@tup.tsinghua.edu.cn
课件下载:https://www.tup.com.cn,010-62795954

印 装 者:三河市龙大印装有限公司
经　　销:全国新华书店
开　　本:185mm×260mm　　印　张:24　　字　数:549千字
版　　次:2010年10月第1版　　2015年6月第2版　　印　次:2024年10月第11次印刷
定　　价:69.00元

产品编号:064110-05

序 言

应用是推动学科技术发展的原动力,计算机科学是实用科学,计算机科学技术广泛而深入地应用推动了计算机学科的飞速发展。应用型创新人才是科技人才的一种类型,应用型创新人才的重要特征是具有强大的系统开发能力和解决实际问题的能力。培养应用型人才的教学理念是教学过程中以培养学生的综合技术应用能力为主线,理论教学以够用为度,所选择的教学方法与手段要有利于培养学生的系统开发能力和解决实际问题的能力。

随着我国经济建设的发展,对计算机软件、计算机网络、信息系统、信息服务和计算机应用技术等专业技术方向的人才的需求日益增加,主要包括软件设计师、软件评测师、网络工程师、信息系统监理师、信息系统管理工程师、数据库系统工程师、多媒体应用设计师、电子商务设计师、嵌入式系统设计师和计算机辅助设计师等。如何构建应用型人才培养的教学体系以及系统框架,是从事计算机教育工作者的责任。为此,中国计算机学会计算机教育专业委员会和清华大学出版社共同组织启动了《中国高等学校计算机科学与技术专业(应用型)学科教程》的项目研究。参加本项目的研究人员全部来自国内高校教学一线具有丰富实践经验的专家和骨干教师。项目组对计算机科学与技术专业应用型学科的培养目标、内容、方法和意义,以及教学大纲和课程体系等进行了较深入、系统的研究,并编写了《中国高等学校计算机科学与技术专业(应用型)学科教程》(简称《学科教程》)。《学科教程》在编写上注意区分应用性人才与其他人才在培养上的不同,注重体现应用型学科的特征。在课程设计中,《学科教程》在依托学科设计的同时,更注意面向行业产业的实际需求。为了更好地体现《学科教程》的思想与内容,我们组织编写了《中国高等学校计算机科学与技术专业(应用型)规划教材》,旨在能为计算机专业应用型教学的课程设置、课程内容以及教学实践起到一个示范作用。本系列教材的主要特点如下:

1. 完全按照《学科教程》的体系组织编写本系列教材,特别是注意在教材设置、教材定位和教材内容的衔接上与《学科教程》保持一致。

2. 每门课程的教材内容都按照《学科教程》中设置的大纲精心编写,尽量体现应用型教材的特点。

3. 由各学校精品课程建设的骨干教师组成作者队伍,以课程研究为基础,将教学的研究成果引入教材中。

4. 在教材建设上,重点突出对计算机应用能力和应用技术的培养,注重教材的实践性。

5. 注重系列教材的立体配套,包括教参、教辅以及配套的教学资源、电子课件等。

高等院校应培养能为社会服务的应用型人才,以满足社会发展的需要。在培养模式、教学大纲、课程体系结构和教材都应适应培养应用型人才的目标。教材体现了培养目标和育人模式,是学科建设的结晶,也是教师水平的标志。本系列教材的作者均是多年从事计算机科学与技术专业教学的教师,在本领域的科学研究与教学中积累了丰富的经验,他们将教学研究和科学研究的成果融入教材中,增强了教材的先进性、实用性和实践性。

目前,我们对于应用型人才培养的模式还处于探索阶段,在教材组织与编写上还会有这样或那样的缺陷,我们将不断完善。同时,也希望广大应用型院校的教师给我们提出更好的建议。

《中国高等学校计算机科学与技术专业(应用型)规划教材》主编

陈明

2008 年 7 月

前 言

STL(Standard Template Library,标准模板库)是 C++ 泛型标准化内容的重要组成部分,主要由容器、迭代器和算法三部分组成,其中封装了数据结构中的绝大部分内容。运用 STL 开发应用程序可以共享各种容器及算法,避免了低层次的各种容器及常用算法的反复开发,在代码一致性、升级、维护等方面都有很大的优越性。因此,学习 STL 是进行深层次开发 C++ 应用程序的重要途径。但是,目前市场上关于 STL 的书籍很多是译著,在思考方法上可能与我们的学生不一致,学习起来很吃力。所以,本书力求把多年的 STL 编程经验按照学生的思维方式进行编排,希望学生们能很快学会 STL 泛型编程方法,体会 STL 泛型编程的乐趣。

本书第 1 版于 2010 年出版,很多读者在使用过程中给予了积极的肯定,并提出了中肯的建议。根据近几年的教学实践,作者对教材内容、开发环境做了调整,使其更适用于不断变化的 C++ 教学及开发。主要修订如下。

(1) 第 1 章~第 10 章所有示例程序,第 11 章部分程序都在 DEV-CPP 编译环境下重新调式,与原来的 VC 稍有不同。很多读者认为 VC 的 STL 有许多漏洞,DEV 更好,支持 gcc 编译。可直接移植到 Linux、Unix 下,所有代码修改的部分均已调试通过。

(2) 删除了 10.11"自定义 STL 风格函数"。该部分涉及 STL 内核程序,而 VC、DEV 下的内核是不同的,这里不宜进行分类讨论。另外,也删除了原 1.3 节"建立 STL 程序的方法"。

(3) 作为配套资源,本书提供所有调试程序的源码,并提供 32 位或 64 位的 DEV 开发环境。

全书共分 11 章,第 1~10 章侧重于基础知识部分,第 11 章侧重于综合应用部分。

第 1 章介绍 STL 的历史和主要内容以及本书用到的开发环境。

第 2 章通过示例说明 STL 中的内存管理思想、重要的 traits 模板技术、模板与操作符重载的关系。

第 3 章介绍 STL 中引入迭代器的原因,并通过自定义迭代器示例加深理解迭代器的内涵。

第 4 章介绍标准输入输出流、文件输入输出流、字符串输入输出流。

第 5 章介绍字符串创建方式及增、删、改、查等常用功能应用方法。

第 6 章介绍引入函数对象的原因,系统函数对象有哪些,自定义函数对象应用方法。

第 7 章介绍 vector、deque、list、queue、stack、priority_queue、bitset、set 和 map 等通用容器的用法，并强调了容器适配器的作用。

第 8～10 章主要是讲算法。第 8 章介绍非变异算法，包括循环、查询、计数、比较等功能；第 9 章介绍变异算法，包含复制、交换、变换、替换、填充、生成、删除、唯一、反转、环移、随机、划分等功能；第 10 章介绍排序及相关操作的算法。

第 11 章侧重于集成应用，包括算法综合应用、在数据结构中应用、在 Visual C++ 中应用三部分。算法综合应用主要介绍在多态、文件解析、综合查询中的 STL 应用方法；在数据结构中应用介绍全排列、频度、最长公共子序列、大整型数加法、乘法、矩阵、回溯、字符串表达式、图中的 STL 应用方法。在 Visual C++ 中应用介绍用 STL 容器存储绘图信息，容器＋算法实现数据保存与查询问题，并介绍 STL 与动态链接库的接口问题等。

本书第 1～5 和第 11 章由刘德山编写，第 6～10 章由金百东编写。因本书程序较多，全书变量均用正体。

本书内容循序渐进，示例丰富，第 1～10 章的所有示例代码编译后就可以运行。第 11 章某些程序由于较大，做了简化处理。示例结果都做了必要的说明，对一些稍难的题目，对其设计思想也做了相应的论述，帮助读者加深对 STL 的理解。

由于作者水平有限，时间紧迫，书中难免有疏漏之处，恳请广大读者批评指正，不胜感激。

编 者

2015 年 3 月

目 录

第1章 STL 概述 1
1.1 STL 历史　1
1.2 STL 内容　2
1.3 命名空间　3

第2章 模板 5
2.1 通过模板初识 STL 思维　5
2.2 traits 技术　8
2.3 模板与操作符重载　12

第3章 迭代器 17
3.1 什么是迭代器　17
3.2 迭代器类位置　22
3.3 进一步理解迭代器　25
3.4 STL 迭代器　26

第4章 输入输出流 31
4.1 标准输入输出流　31
4.1.1 插入符与提取符　31
4.1.2 get 系列函数　33
4.1.3 处理流错误　34
4.2 文件输入输出流　36
4.2.1 文件打开　36
4.2.2 文件关闭　36
4.2.3 文件读写　36
4.3 字符串输入输出流　41
4.4 综合示例　42

第5章 字符串 ··· 47

5.1 字符串创建及初始化 47
5.1.1 基本创建方式 47
5.1.2 迭代器创建方式 48
5.2 字符串操作 48
5.2.1 插入操作 48
5.2.2 替换操作 49
5.3 字符串查询 50
5.4 在字符串中删除字符 52
5.5 字符串比较 52
5.6 综合示例 53

第6章 函数对象 ··· 59

6.1 简介 59
6.1.1 为何引入函数对象 59
6.1.2 函数对象分类 60
6.1.3 简单示例 61
6.2 一元函数 62
6.3 二元函数 64
6.4 系统函数对象 66
6.4.1 算术类函数对象 67
6.4.2 关系运算类函数对象 70
6.4.3 逻辑运算类函数对象 72
6.4.4 函数适配器 72
6.5 综合示例 77

第7章 通用容器 ··· 81

7.1 概述 81
7.1.1 容器分类 81
7.1.2 容器共性 82
7.1.3 容器比较 83
7.2 vector容器 83
7.2.1 概述 83
7.2.2 初始化示例 84
7.2.3 增加及获得元素示例 86
7.2.4 修改元素示例 90

7.2.5　删除元素示例　91
　　　7.2.6　进一步理解 vector　92
　　　7.2.7　综合操作示例　93
　7.3　deque 容器　97
　　　7.3.1　常用函数　97
　　　7.3.2　基本操作示例　98
　　　7.3.3　综合操作示例　100
　7.4　list 容器　102
　　　7.4.1　常用函数　103
　　　7.4.2　基本操作示例　104
　　　7.4.3　综合操作示例　107
　7.5　队列和堆栈　113
　　　7.5.1　常用函数　113
　　　7.5.2　容器配接器　114
　　　7.5.3　基本操作示例　115
　　　7.5.4　综合操作示例　118
　7.6　优先队列　121
　　　7.6.1　常用函数　121
　　　7.6.2　基本操作示例　122
　　　7.6.3　综合操作示例　123
　7.7　bitset 容器　126
　　　7.7.1　常用函数　126
　　　7.7.2　基本操作示例　127
　　　7.7.3　综合操作示例　130
　7.8　集合　133
　　　7.8.1　常用函数　133
　　　7.8.2　基本操作示例　134
　　　7.8.3　综合操作示例　137
　7.9　映射　140
　　　7.9.1　常用函数　140
　　　7.9.2　基本操作示例　141
　　　7.9.3　综合操作示例　144
　7.10　再论迭代器　148

第8章　非变异算法　153

　8.1　循环　153
　　　8.1.1　主要函数　153
　　　8.1.2　示例分析　154

8.2 查询 **158**
    8.2.1 主要函数 **158**
    8.2.2 示例分析 **161**
8.3 计数 **169**
    8.3.1 主要函数 **169**
    8.3.2 示例分析 **170**
8.4 比较 **172**
    8.4.1 主要函数 **172**
    8.4.2 示例分析 **173**

第9章 变异算法 *177*

9.1 复制 **178**
    9.1.1 主要函数 **178**
    9.1.2 示例分析 **179**
9.2 交换 **180**
    9.2.1 主要函数 **180**
    9.2.2 示例分析 **181**
9.3 变换 **182**
    9.3.1 主要函数 **182**
    9.3.2 示例分析 **183**
9.4 替换 **186**
    9.4.1 主要函数 **186**
    9.4.2 示例分析 **188**
9.5 填充 **190**
    9.5.1 主要函数 **190**
    9.5.2 示例分析 **191**
9.6 生成 **192**
    9.6.1 主要函数 **192**
    9.6.2 示例分析 **193**
9.7 删除 **198**
    9.7.1 主要函数 **198**
    9.7.2 示例分析 **199**
9.8 唯一 **204**
    9.8.1 主要函数 **204**
    9.8.2 示例分析 **205**
9.9 反转 **207**
    9.9.1 主要函数 **207**

9.9.2 示例分析 208
9.10 环移 209
　9.10.1 主要函数 209
　9.10.2 示例分析 210
9.11 随机 212
　9.11.1 主要函数 212
　9.11.2 示例分析 212
9.12 划分 215
　9.12.1 主要函数 215
　9.12.2 示例分析 216

第10章 排序及相关操作 ……………………………………………………… 219

10.1 排序 220
　10.1.1 主要函数 220
　10.1.2 示例分析 222
10.2 第n个元素 227
　10.2.1 主要函数 227
　10.2.2 示例分析 228
10.3 二分检索 229
　10.3.1 主要函数 229
　10.3.2 示例分析 231
10.4 归并 232
　10.4.1 主要函数 232
　10.4.2 示例分析 233
10.5 有序结构上的集合操作 234
　10.5.1 主要函数 234
　10.5.2 示例分析 237
10.6 堆操作 242
　10.6.1 主要函数 242
　10.6.2 示例分析 244
10.7 最大和最小 247
　10.7.1 主要函数 247
　10.7.2 示例分析 248
10.8 词典比较 249
　10.8.1 主要函数 249
　10.8.2 示例分析 250
10.9 排列生成器 251

10.9.1 主要函数 251
10.9.2 示例分析 252
10.10 数值算法 253
10.10.1 主要函数 253
10.10.2 示例分析 255

第 11 章 STL 应用 257

11.1 算法的综合运用 257
11.1.1 在多态中的应用 257
11.1.2 set、map 应用 261
11.1.3 ini 文件解析 264
11.1.4 综合查询 269
11.2 在数据结构中的应用 280
11.2.1 全排列应用 280
11.2.2 频度问题 283
11.2.3 最长公共子序列问题 285
11.2.4 大整型数加法、乘法类 288
11.2.5 矩阵问题 293
11.2.6 回溯问题 296
11.2.7 字符串表达式 300
11.2.8 图 306
11.3 在 Visual C++ 中应用 316
11.3.1 Scribble 绘图程序 317
11.3.2 数据库操作程序 324
11.3.3 文本文件排序、查询 337
11.3.4 基于配置文件的查询程序 346
11.3.5 STL 与动态链接库 360

参考文献 369

Chapter 1

第 1 章　STL 概述

1.1　STL 历史

20 世纪 70 年代,被誉为标准模板库(Standard Template Library,STL)之父的 Alexander Stepanov 开始考虑:在保证效率的前提下,将算法从诸多具体应用之中抽象出来的可能性。为了验证自己的思想,他和纽约州立大学教授 Deepak Kapur,伦塞里尔技术学院教授 David Musser 共同开发了一种叫做 Tecton 的语言。尽管这次尝试没有取得实用性的成果,但却给了 Stepanov 很大的启示。

在随后的几年中,他又和 David Musser 等人先后用 Schema 语言(一种 Lisp 语言的变种)和 Ada 语言建立了一些大型程序库。Alexander Stepanov 逐渐意识到:在当时的面向对象程序设计思想中存在一些问题,比如抽象数据类型概念所存在的缺陷。Stepanov 希望通过对软件领域中各组成部分的分类,逐渐形成一种软件设计的概念性框架。

1987 年,在贝尔实验室工作的 Alexander Stepanov 开始首次采用 C++ 语言进行泛型软件库的研究。由于当时的 C++ 语言还没有引入模板(template)语法,研究方法只能采用继承机制。尽管如此,Stepanov 还是开发出了一个庞大的算法库。与此同时,在与 Andrew Koenig(前 ISO C++ 标准化委员会主席)和 Bjarne Stroustrup(C++ 语言的创始人)等顶级大师们的共事过程中,Stepanov 开始注意到 C/C++ 语言在实现其泛型思想方面所具有的潜在优势。就拿 C/C++ 中的指针而言,它的灵活与高效运用使后来的 STL 在实现泛型化的同时更是保持了高效率。另外,在 STL 中占据极其重要地位的迭代子概念便是源自于 C/C++ 中原生指针(native pointer)的抽象。

1988 年,Alexander Stepanov 开始进入惠普的 Palo Alto 实验室工作,在随后的 4 年中,他从事的是有关磁盘驱动器方面的工作。直到 1992 年,由于参加并主持了实验室主任 Bill Worley 所建立的一个有关算法的研究项目,才使他重新回到了泛型化算法的研究工作上来。项目自建立之后,参与者从最初的 8 人逐渐减少,最后只剩下两个人——Stepanov 和 Meng Lee。经过长时间的努力,最终完成了一个包含有大量数据结构和算法部件的庞大运行库。这便是现在 STL 的雏形。

1993年,当时在贝尔实验室的Andrew Koenig看到了Stepanov的研究成果,在他的鼓励与帮助下,Stepanov于1993年9月在圣何塞为ANSI/ISO C++标准委员会做了一个相关演讲(题为"The Science of C++ Programming"),向委员们讲述了其观念。然后又于1994年3月,在圣迭戈会议上向委员会提交了一份建议书,以期使STL成为C++标准库的一部分。尽管这一建议十分庞大,以至于降低了被通过的可能性,但其所包含的新思想吸引了许多人的注意力。

随后,在众人的帮助之下(包括Bjarne Stroustrup),Stepanov又对STL进行了改进。同时加入了一个封装内存模式信息的抽象模块,也就是现在STL中的allocator,它使STL的大部分实现都可以独立于具体的内存模式,从而独立于具体平台。在1994年的滑铁卢会议上,委员们最终通过了提案,决定将STL正式纳入C++标准化进程之中,随后STL便被放进了会议的工作文件中。自此,STL终于成为C++家族中的重要一员。

此后,随着C++标准的不断改进,STL也在不断地做着相应的演化。直至1998年,ANSI/ISO C++标准正式定案,STL始终是C++标准中不可或缺的一大部件。

1.2 STL内容

STL主要包含容器、算法和迭代器三大部分。

STL容器包含了绝大多数数据结构,如数组、链表、队列、堆、栈和树等。开发者直接应用这些系统STL容器相关函数就可以了,而且这些函数都是带模板参数的,可以适应许多数据元素类型,功能非常强大。STL算法包含了诸如增、删、改、查和排序等系统函数,开发者可以直接操作这些函数实现相应功能。STL迭代器类似指针,通过它的有序移动把容器中的元素与算法关联起来,它是实现所有STL功能的基础所在。

当然,STL也包含其他一些内容,如字符串、输入输出流等内容。

本书主要是以DEV-CPP中内嵌的STL为基础讲述的,开发环境选择DEV,第11章部分示例用到了Visual C++ 6.0。用到的主要头文件如表1.1所示。

表1.1 常用STL包含文件

索引	功能	包含文件	备注
1	迭代器	#include <iterator>	
2	输入输出流	#include <iostream>	标准输入输出流
		#include <fstream>	文件输入输出流
		#include <sstream>	字符串输入输出流
3	字符串	#include <string>	
4	函数对象	#include <functional>	

续表

索引	功能	包含文件	备注
5	通用容器	#include <vector>	向量容器
		#include <deque>	双端队列
		#include <list>	链表容器
		#include <query>	队列、优先队列
		#include <stack>	堆栈
		#include <set>	集合、多集合、位集合
		#include <map>	映射、多映射
6	通用算法	#include <algorithm>	
7	数值算法	#include <numeric>	

STL 的包含文件都不加扩展名，以便与 C 语言风格的".h"头文件相区别。

1.3 命名空间

STL 程序中一般需要加"using namespace std;"，这属于命名空间的应用,关键字是 namespace。在 C++ 中，名称可以是变量、函数、结构、枚举以及类和结构的成员。随着工程的增大，名称相互冲突的可能性也将增加。使用多个厂商的类库时，可能导致名称冲突。例如两个库可能都定义了名为 func 的函数，但定义的方式不兼容，那么在应用中如何应用某一个具体的 func 函数呢？C++ 提出了 namespace 命名空间解决方法，只要再加一层封装就可以了。

【例 1.1】 namespace 命名空间示例。

```
//文件名：e1_1.cpp
#include <stdio.h>
namespace mycompany
{
    void func()
    {
        printf("Hello, this is my company");
    }
};
namespace yourcompany
{
    void func()
    {
```

```
        printf("Hello, this is your company");
    }
};
using namespace mycompany;
int main(int argc, char * argv[])
{
    func();
    printf("\n");
    yourcompany::func();
    printf("\n");
    return 0;
}
```

由于 STL 的命名空间名称是 std，因此要应用 STL，必须要包含"using namespace std;"。

第 2 章 模 板

模板是 C++ 语言中重要的概念。它提供了一种通用的方法来开发可重用的代码,即可以创建参数化的 C++ 类型。模板分为两种类型:函数模板和类模板。函数模板的用法同 C++ 预处理器的用法有一定的类似之处,它们都提供编译代码过程中的文本替换功能,但前者可以对类型进行一定的保护。使用类模板可以编写通用的、类型安全的类。STL 中仍然要用到函数模板和类模板,本章旨在让学生们理解一些具体的 STL 模板思维。

2.1 通过模板初识 STL 思维

学过 C++ 的人都学过模板技术,那么是否与 STL 模板不一样呢?其实 STL 并不是特别神秘,用到的也不是新技术,下面通过例子加以说明。

【例 2.1】 编制动态数组的模板类。

```
//文件名:e2_1.cpp
#include<stdio.h>
template<class T>
class MyArray
{
private:
    int m_nTotalSize;                               //数组总长度
    int m_nValidSize;                               //数组有效长度
    T * m_pData;                                    //数据
public:
    MyArray(int nSize=3)                            //数组默认总长度是 3
    {
        m_pData=new T[nSize];
        m_nTotalSize=nSize;
        m_nValidSize=0;
    }
    void Add(T value)                               //向 m_pData 添加数据
    {
        if(m_nValidSize<m_nTotalSize)               //如果有效长度小于总长度
```

```cpp
        {
            m_pData[m_nValidSize]=value;        //则赋值
            m_nValidSize++;                     //有效长度+1
        }
        else                                    //如果有效长度大于等于总长度
        {
            int i=0;
            T * tmpData=new T[m_nTotalSize];    //原始数据备份
            for(i=0; i<m_nTotalSize; i++)
            {
                tmpData[i]=m_pData[i];
            }
            delete []m_pData;                   //释放原始数据内存空间
            m_nTotalSize *= 2;                  //原始数据空间重新分配,空间扩大1倍
            m_pData=new T[m_nTotalSize];        //传回备份数据
            for(i=0; i<m_nValidSize; i++)
            {
                m_pData[i]=tmpData[i];
            }
            delete []tmpData;
            m_pData[m_nValidSize]=value;
            m_nValidSize++;
        }
    }
    int GetSize()                               //返回数组有效长度
    {
        return m_nValidSize;
    }
    T Get(int pos)                              //返回某一位置元素
    {
        return m_pData[pos];
    }
    virtual~MyArray()
    {
        if(m_pData!=NULL)
        {
            delete []m_pData;
            m_pData=NULL;
        }
    }
};
int main(int argc, char * argv[])
```

```
    MyArray<int>obj;
    obj.Add(1);
    obj.Add(2);
    obj.Add(3);
    obj.Add(4);
    for(int i=0; i<obj.GetSize(); i++)
    {
        printf("%d\n", obj.Get(i));
    }
    return 0;
}
```

（1）本示例是一个简单的动态数组模板类，在 main 函数中，先通过调用 MyArray 构造函数产生了默认大小为 3 个元素的数组指针 m_pData，然后通过 Add 函数向其中增加了 4 个元素，最后完成了动态数组元素的显示。m_pData 仅包含 3 个元素空间，却添加了 4 个元素，仔细分析代码 Add 函数中黑体部分内容，就可以明白其中的道理：当添加前 3 个元素时，m_pData 内存空间没有变化；当添加第 4 个元素（obj.Add(4)）时，m_pData 已经没有多余空间容纳数据了，则把 m_pData 中数据先保存在临时数据 tmpData 中。销毁 m_pData，重新分配 m_pData 内存空间，大小是原先的 2 倍，由 3 变成 6。之后把临时数据 tmpData 中的数据再复制回 m_pData 中，最后把第 4 个元素（obj.Add(4)）加到 m_pData 中。可以推之：当加到第 7 个元素时，m_pData 空间由 6 变成 12；当加到第 13 个元素时，m_pData 空间由 12 变成 24……也就是说，通过黑体代码，调用方可随意添加元素，实现了动态数组的生成。

（2）本例能体现出 STL 容器关于内存"动态分配、销毁、再分配"的思想，也就是把与内存管理的部分进一步抽象，编成系统代码，应用方不必明了过程中的内存变化，用专家级编制的代码，而不是自己编制的代码来管理内存。普通人编制的代码不如 STL 专家编制得专业，STL 是编制普通模板类发展的必然结果，不是一种新技术。

【例 2.2】 编制一个数组元素求和的函数模板。

为了更好地说明问题，表 2.1 中列出了整型数组元素求和函数与题中要求的函数模板的代码对比。

表 2.1 非模板函数与模板函数代码对比

非模板函数（整型数组元素和）	模板函数（任意类型数组元素和）

`int sum(int data[], int nSize)`
`{`
    `int sum=0;`
    `for(int i=0; i<nSize; i++)`
        `sum+=data[i];`
    `return sum;`
`}`
	`template <class T>`
	`T sum(T data[], int nSize)`
	`{`
	    `T sum=0;`
	    `for(int i=0; i<nSize; i++)`
	        `sum+=data[i];`
	    `return sum;`
	`}`

(1) 表中非模板函数转化成模板函数可以这样来分析：增加一行模板参数声明 template <class T>，再把所有的 int 用 T 来代替即可。因此函数模板并不难编制，如果仍然感到困难，不妨先编制一个关于基本数据类型的非模板函数，再用 T 来代替某基本数据类型的声明，那么离实际的泛型模板函数也就很近了。

(2) 从非模板函数到模板函数，这在编程思维上已经有了很大的飞跃。但是该模板函数还有一定的局限性，那就是它只能对数组元素求和。那么进一步思考，希望用该函数仍能实现对链表、集合等元素的求和，这即是 STL 的思维方式。如果想到了这点，并加以编码实现，或许你编制的代码与 STL 源码有很大的相似性，这对于理解 STL 有很大的好处。希望同学们课下去实践。

2.2 traits 技术

STL 标准模板库非常强调软件的复用，traits 技术是重要的手段。traits 的中文意思就是特性，traits 就像特性萃取机，提取不同类的共性，以便能统一处理。traits 依靠显式模板特殊化来把代码中因类型不同而发生变化的片段拖出来，用统一的接口来包装。这个接口可以包含一个 C++ 类所能包含的任何东西，如内嵌类型、成员函数、成员变量。作为客户的模板代码，可以通过 traits 模板类所公开的接口来间接访问。下面通过一个简单的例子加以理解。

【例 2.3】 已知整型数组类 CIntArray，浮点数组类 CFloatArray，求整型或浮点数组的和乘以相应倍数并输出。

代码如下所示。

```
//文件名：e2_3.cpp
#include <iostream>
using namespace std;
class CIntArray
{
    int a[10];
public:
    CIntArray()                                    //a[0]=1,a[1]=2,…
    {
        for(int i=0; i<10; i++)
        {
            a[i]=i+1;
        }
    }
    int GetSum(int times)                          //times：整数倍数
    {
        int sum=0;
```

```cpp
            for(int i=0; i<10; i++)
            {
                sum+=a[i];
            }
            return sum * times;
        }
    };

    class CFloatArray
    {
        float f[10];
    public:
        CFloatArray()                              //a[0]=1,a[1]=1/2,…
        {
            for(int i=1; i<=10; i++)
            {
                f[i-1]=1.0f/i;
            }
        }
        float GetSum(float times)                  //浮点倍数
        {
            float sum=0.0f;
            for(int i=0; i<10; i++)
            {
                sum+=f[i];
            }
            return sum * times;
        }
    };
    int main()
    {
        CIntArray intary;
        CFloatArray fltary;
        cout<<"整型数组和 3 倍是:"<<intary.GetSum(3)<<endl;
        cout<<"浮点数组和 3.2 倍是:"<<fltary.GetSum(3.2f)<<endl;
        return 0;
    }
```

很简单的功能,即把整型或浮点数组的和乘以相应倍数并输出。CIntArray、CFloatArray 功能相似,main 函数中通过调用相应类(两个类)的 GetSum 函数完成。那么能否通过一个类的接口函数来完成上述功能呢?可以,当然要用到模板,增加一个类 CApply。

```cpp
template<class T>
class CApply
{
public:
    float GetSum(T& t, float inpara)
    {
        return t.GetSum(inpara);
    }
};
```

main 函数变为如下代码：

```cpp
void main()
{
    CIntArray intary;
    CFloatArray fltary;
    CApply<CIntArray>c1;
    CApply<CFloatArray>c2;
    cout<<"整型数组和 3 倍是:" <<c1.GetSum(intary, 3)<<endl;
    cout<<"浮点数组和 3.2 倍是:" <<c2.GetSum(fltary, 3.2f)<<endl;
}
```

通过模板类 CApply 接口函数实现了对整型数组及浮点数组类的操作。但是仔细分析一下，细节处还是有问题的：比如 CIntArray 类中 GetSum 函数返回值是整数，函数输入参数是整型；CFloatArray 类中 GetSum 函数返回值是浮点类型，函数输入参数是浮点类型。而模板类 CApply 中 GetSum 把返回值固定成 float，输入参数也固定成 float，虽然从结果看正确，但是从严密角度来说不够严密，如果问题复杂了，结果可能就不正确了。那么如何解决输入、输出参数类型的不同呢？traits 技术就是很好的解决方法。步骤如下所示。

（1）定义基本模板类。

```cpp
template<class T>
class NumTraits
{
};
```

NumTraits 可以什么内容都不写，只是说明该类是一个模板类。

（2）模板特化。本示例是针对 CIntArray 及 CFloatArray 的，因此代码如下所示。

```cpp
template<>
class NumTraits<CIntArray>
{
public:
    typedef int resulttype;
    typedef int inputpara;
```

```
    template<>
    class NumTraits<CFloatArray>
    {
    public:
        typedef float resulttype;
        typedef float inputpara;
    };
```

可以看出相应模板特化类中只是用了 typedef 重定义函数。NumTraits<CIntArray>中，根据 CIntArray 类中函数 int GetSum(int times)，把返回值 int 类型重新定义成 resulttype，把输入 int 类型重新定义成 inputpara。同理，NumTraits<CFloatArray>中也定义了相应的 resulttype 及 inputpara。那么为什么要把返回类型、输入参数都定义成相同的名称呢？它是为编制模板类共同的调用接口做准备。

(3) 统一模板调用类编制。

```
    template<class T>
    class CApply
    {
    public:
        NumTraits<T>::resulttype GetSum(T& obj, NumTraits<T>::inputpara in)
        {
            return obj.GetSum(in);
        }
    };
```

乍一看起来，GetSum 函数有些难懂。当模板参数代表 CIntArray，该定义变为如下代码：

```
    NumTraits<CIntArray>::resulttype GetSum(CIntArray& obj, NumTraits<CIntArray>::
    inputpara in)
```

根据(2)中模板特化定义，NumTraits<CIntArray>::resulttype 代表 int，NumTraits<CIntArray>::inputpara 代表 int，于是上式变为 int GetSum(CIntArray& obj, int in)。

当模板参数代表 CFloatArray，该定义变为如下代码：

```
    NumTraits<CFloatArray>::resulttype GetSum(CFloatArray& obj, NumTraits<CFloatArray
    >::inputpara in)
```

根据(2)中模板特化定义，NumTraits<CFloatArray>::resulttype 代表 int，NumTraits<CFloatArray>::inputpara 代表 int，于是上式变为 float GetSum(CIntArray& obj, float in)。

因此原函数 NumTraits<T>::resulttype GetSum(T& obj, NumTraits<T>::inputpara in)中返回值及输入参数的类型是可变的，随 NumTraits<T>::resulttype、

NumTraits<T>::inputpara 变化而变化。所以在模板特化类中给输入、输出参数进行 typedef 重定义非常重要,而且起的对应名称还要相同。

为了简化 CApply 函数中定义形式,再次巧妙运用 typedef 进行定义,代码如下所示,与原始功能是相同的。

```
template<class T>
class CApply
{
public:
    typedef typename NumTraits<T>::resulttype result;
    typedef typename NumTraits<T>::inputpara input;
    result GetSum(T& obj, input in)
    {
        return obj.GetSum(in);
    }
};
```

2.3 模板与操作符重载

例如有关于大小的模板函数,代码如下所示。

```
template<class U, class V>
bool MyGreater(U& u, V& v)
{
    return u>v;
}
```

这是两种类型数据泛型比较程序,在用以下代码,即基本数据类型做参数时,程序都正确。

```
MyGreater(1,2);
MyGreater(1.5,2.5);
MyGreater('a', 10);
```

但是当 U 或 V 有一个表示类时,编译通不过,例如已知学生基本类代码如下所示。

```
class Student
{
    char name[20];                    //姓名
    int grade;                         //成绩
public:
    Student(char name[], int grade)
    {
```

```
        strcpy(this->name, name); this->grade=grade;
    }
};
```

当想获得某学生成绩是否大于 75 分时,或许希望写成如下形式。

```
Student stud("zhang san", 80);                    //定义学生对象
MyGreater(stud, 75);                              //学生成绩是否大于 75
```

当调用 MyGreater 模板函数时相当于如下代码。

```
bool MyGreater(Student& u, int v)
{
    return u>v;
}
```

u 是 Student 对象,v 是基本数据类型,两者不能直接比较,必须重载类中操作符。本例中由于是 u>v,因此必须重载 Student 类中">"运算符;由于 v 是整型数,因此操作符传入参数类型是整型数;又由于仅是比较大小,返回值是 bool。这样,Student 类变为如下代码。

```
class Student
{
    char name[20];                                //姓名
    int grade;                                    //成绩
public:
    Student(char name[], int grade)
    {
        strcpy(this->name, name); this->grade=grade;
    }
    bool operator> (const int &value)const
    {
        return grade>value;
    }
};
```

再如一个关于数组二分查找的模板函数,前提是数组按某规则已经排好序,代码如下所示。

```
template<class T1,class T2>
int Binary_find(T1 t[], int nSize,T2 value)       //nSize 数组大小
{                                                 //value:二分查找数值
    int left=0;
    int right=nSize-1;
    int mid=0;
    bool bFind=false;
    while(1&&(right>=left))
```

```
        {
            mid=(left+right)/2;
            if(t[mid]==value)----------------(a)
            {
                bFind=true;
                break;
            }
            if(t[mid]<value)----------------(b)
            {
                left=mid+1;
            }
            if(t[mid]>value)----------------(c)
            {
                right=mid-1;
            }
        }
        int pos=bFind==true?mid:-1;
        return pos;           //pos:找到value元素的数组下标,当pos=-1时,表示没有找到
    }
```

对基本数据类型而言,直接编译执行就可以了,如下述测试代码。

```
int a[]={1,3,5,7,9,11,13,15,17,19};
int pos1=Binary_find(a, 10, 13);
int pos2=Binary_find(a, 10, 8);
```

但如果对类对象而言,就要考虑重载(a)、(b)、(c)处的操作符。例如,学生对象数组已经按成绩升序排列,现在想查询一下有无成绩是70分的学生,测试代码想写成:

```
Student s[]=…;                    //初始化完毕
Binary_find(s, 10, 70);           //查询10个学生集合s[]中有无成绩70分的学生
```

这样,可以得出,模板参数T1对应Student,模板参数T2对应int,因此Student类应重载operate==、operate>、operate<操作符,代码如下所示。

```
class Student
{
    char name[20];                                //姓名
    int grade;                                    //成绩
public:
    Student(char name[], int grade)
    {
        strcpy(this->name, name); this->grade=grade;
    }
```

```
    bool operator>(const int &value)const
    {
        return grade>value;
    }
    bool operate<(const int &value)const
    {
        return grade<value;
    }
    bool operate==(const int &value)const
    {
        return grade==value;
    }
};
```

由于 STL 中有大量的模板函数,因此很多时候要重载与之对应的操作符。模板函数相当于已编制好的应用框架,操作符重载相当于调用的接口。

第 3 章 迭 代 器

迭代器是 STL 的核心技术，提供了统一访问容器元素的方法，为编制通用算法提供了坚实的技术基础。

3.1 什么是迭代器

迭代器即指针，可以是所需要的任意类型，它的最大好处是可以使容器和算法分离。例如，有两个容器类：MyArray 是某类型数组集合；MyLink 是某类型链表集合。它们都有显示、查询和排序等功能，常规思维是每个容器类中有自己的显示、查询和排序等函数。仔细分析可得出：不同容器中完成相同功能代码的思路大体是相同的，那么能不能把它们抽象出来，多个容器仅对应一个显示、一个查询、一个排序函数呢？这是泛型思维发展的必然结果，于是迭代器思维就产生了。下面通过示例加深对迭代器的理解。

【例 3.1】 为数组容器、链表容器编制共同显示函数。

MyArray 几乎与例 2.1 中的相同，如下所示。

```
//文件名：e3_1.cpp(本示例中所有头文件及源文件内容都在该文件中)
#include<stdio.h>
template<class T>
class MyArray
{
private:
    int m_nTotalSize;                    //数组总长度
    int m_nValidSize;                    //数组有效长度
    T * m_pData;                         //数据
public:
    MyArray(int nSize=3)                 //数组默认总长度是 3
    {
        m_pData=new T[nSize];
        m_nTotalSize=nSize;
        m_nValidSize=0;
    }
    void Add(T value)                    //向 m_pData 添加数据
```

```cpp
    {
        ...                                    //同例 2.1
    }
    int GetSize()                              //返回数组有效长度
    {
        return m_nValidSize;
    }
    T Get(int pos)                             //返回某一位置元素
    {
        return m_pData[pos];
    }
    virtual ~MyArray()
    {
        if(m_pData!=NULL)
        {
            delete []m_pData;
            m_pData=NULL;
        }
    }
};
```

MyLink 单项链表类初始代码如下所示。

```cpp
template <class T>
struct Unit                                    //链表单元
{
    T value;
    Unit * next;
};
template <class T>
class MyLink
{
    Unit<T> * head;                            //链表头
    Unit<T> * tail;                            //链表尾
    Unit<T> * prev;
public:
    MyLink()
    {
        head=tail=prev=NULL;
    }
    void Add(T &value)                         //向链表中添加元素
    {
        Unit<T> * u=new Unit<T>();
        u->value=value;
```

```
            u->next=NULL;
        if(head ==NULL)
        {
            head=u;
            prev=u;
        }
        else
        {
            prev->next=u;
            prev=u;
        }
        tail=u->next;
    }
    virtual~MyLink()
    {
        if(head!=NULL)
        {
            Unit<T> * prev=head;
            Unit<T> * next=NULL;
            while(prev!=tail)
            {
                next=prev->next;
                delete prev;
                prev=next;
            }
        }
    }
};
```

可以看出，MyLink 是模板元素 T 的链表类，以 struct Unit 为一个个链表单元。那么如何以 MyArray、MyLink 为基础完成一个共同显示函数呢？其实非常简单，先从需要出发，逆向考虑，写一个泛型显示函数，如下所示。

```
template<class Init>
void display(Init start, Init end)
{
    cout<<endl;
    for(Init mid=start; mid!=end; mid++)
    {
        cout<< * mid<<"\t";
    }
    cout<<end;
}
```

模板参数类型 Init 是一个指针,start 是起始指针,end 是结束指针,指针支持＋＋及＊操作。display 函数与具体的容器无直接关联,间接关联是必需的。对于 MyArray 来说,Init 相当于对 T＊的操作;对于 MyLink 来说,Init 相当于对 Unit<T>＊的操作。因此就引出了迭代器类,它仍然是一个模板类,对于本示例而言,模板参数是 T 或 Unit<T>＊。

MyArray 对应的迭代器 ArrayIterator 类为:

```
template<class Init>
class ArrayIterator
{
    Init * init;
public:
    ArrayIterator(Init * init)
    {
        this->init=init;
    }
    bool operator!=(ArrayIterator& it)
    {
        return this->init!=it.init;
    }
    void operator++(int)
    {
        init++;
    }
    Init operator * ()
    {
        return * init;
    }
};
```

可以看出,ArrayIterator 确实是对 Init＊的再封装,必须重载!＝、＋＋、＊操作符,这是由于 display 泛型显示函数用到这些操作符的缘故。

另外,还需要在 MyArray 类中增加两个函数 Begin、End,用以获得起止迭代指针,如下所示。

```
T * Begin()                          //起始迭代指针
{
    return m_pData;
}

T * End()                            //结束迭代指针
{
    return m_pData+m_nValidSize;
}
```

测试函数如下所示。

```
int main()
{
    MyArray<int>ary;
    for(int i=0; i<5; i++)
    {
        ary.Add(i+1);
    }
    ArrayIterator<int>start(ary.Begin());
    ArrayIterator<int>end(ary.End());
    cout<<"数组元素为:";
    display(start, end);
    return 0;
}
```

过程可描述为：定义一个 MyArray 对象 ary，向其中添加了 5 个数据(1～5)，获得起止迭代指针对象 start、end，最后调用泛型显示函数 display 完成数据的输出。

同理，可知完成 MyLink 迭代功能步骤如下所示。

(1) 在 MyLink 类中增加 Begin()、End()，用以获取起止迭代指针。

```
Unit<T> * Begin()                              //链表头指针
{
    return head;
}
Unit<T> * End()                                //链表尾指针
{
    return tail;
}
```

(2) 增加链表迭代器类 LinkIterator，重载!=、++、* 运算符。

```
template<class Init>
class LinkIterator
{
    Init * init;
public:
    LinkIterator(Init * init)
    {
        this->init=init;
    }
    bool operator!=(LinkIterator& it)
    {
        return this->init!=it.init;
    }
    void operator++(int)
    {
        init=init->next;
```

```
    }
    Init operator * ()
    {
        return * init;
    }
};
```

可以看出 operator ！＝、operator ＊ 重载内容与 ArrayIterator 中的内容是相同的,只有 operator ＋＋ 中的内容不同。对于链表而言,不像数组那样内存是连续的,是指针的转向,因此绝对不能写成 init＝init＋＋,只能写成 init＝init－＞next。

（3）重载全局函数 operator ＜＜。

```
template<class T>
ostream& operator<<(ostream& os, Unit<T>& s)
{
    os<<s.value;
    return os;
}
```

这是因为当执行 display 函数中的 cout＜＜＊mid 指令时,对链表而言表意形式相当于 cout＜＜＊Unit＜int＞,即输出 Unit＜int＞对象。由于 Unit 是复合数据类型,不能直接输出,因此必须重载上述函数,在其中输出简单数据类型 s.value。

（4）链表类测试函数如下：

```
int main()
{
    int m=0;
    MyLink<int>ml;
    for(int i=0; i<5; i++)
    {
        m=i+1;
        ml.Add(m);
    }
    LinkIterator<Unit<int>>start(ml.Begin());
    LinkIterator<Unit<int>>end(ml.End());
    display(start, end);
    retrun 0;
}
```

3.2 迭代器类位置

3.1 节中描述了自定义数组迭代器、链表迭代器的编制方法,迭代器类位于容器之外。但仔细分析能够发现：数组迭代器只能应用于数组容器,不能应用于链表容器；链表迭代器只能应用于链表容器,不能应用于数组容器。也就是说,特定的容器应该有特定的迭代器。因此,把迭代器类作为容器的内部类更符合应用的特点。以 MyLink 为例,融合 LinkIterator 后代码如下(struct Unit 结构体也是隶属于 MyLink 容器的,因此也把它一起合并):

```cpp
template<class T>
class MyLink
{
public:
    struct Unit                                    //链表单元
    {
        T value;
        Unit * next;
    };
    class LinkIterator
    {
        Unit * init;
    public:
        LinkIterator(Unit * init)
        {
            this->init=init;
        }
        bool operator!=(LinkIterator& it)
        {
            return this->init!=it.init;
        }
        void operator++(int)
        {
            init=init->next;
        }
        Unit operator * ()
        {
            return * init;
        }
    };
    Unit * head;                                   //链表头
    Unit * tail;                                   //链表尾
    Unit * prev;
public:
    MyLink()
    {
        head=tail=prev=NULL;
    }
```

```cpp
    void Add(T &value)                    //向链表中添加元素
    {
        Unit * u=new Unit();
        u->value=value;
        u->next=NULL;
        if(head==NULL)
        {
            head=u;
            prev=u;
        }
        else
        {
            prev->next=u;
            prev=u;
        }
        tail=u->next;
    }
    Unit * Begin()
    {
        return head;
    }
    Unit * End()
    {
        return tail;
    }
    virtual~MyLink()
    {
        if(head!=NULL)
        {
            Unit * prev=head;
            Unit * next=NULL;
            while(prev!=tail)
            {
                next=prev->next;
                delete prev;
                prev=next;
            }
        }
    }
};
```

与原始代码相比简洁许多,少了许多尖括号。其他需要稍许变化的地方及测试代码如下所示。

```
template<class T>
ostream& operator<<(ostream& os, MyLink<T>::Unit& s)
//由 Unit<T>& s 变为 MyLink<T>::Unit& s
{
    os<<s.value;
    return os;
}
int main()
{
    int m=0;
    MyLink<int>ml;
    for(int i=0; i<5; i++)
    {
        m=i+1;
        ml.Add(m);
    }
    MyLink<int>::LinkIterator start=obj2.Begin();
    MyLink<int>::LinkIterator end=obj2.End();
    display(start, end);
    return 0;
}
```

3.3 进一步理解迭代器

3.1节和3.2节详细讨论了迭代器的编程思路,如果用图形来描述则如图3.1所示。

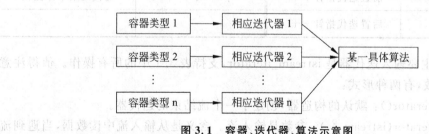

图 3.1 容器、迭代器、算法示意图

每个容器都应有对应的迭代器,容器通过迭代器共享某一具体算法,某一具体算法不依附于某一具体的容器。迭代器起到一个中间媒介的作用,通过它把容器与算法关联起来。换一句更贴切的话来说,迭代器思维是编制通用泛型算法发展的必然结果,算法通过迭代器依次访问容器中的元素。

由图3.1也可以得出STL标准模板库编程的基本步骤:
① 形成容器元素;

② 取出所需要的迭代指针；

③ 调用通用算法。

其实，在生活中有许多"迭代器"的现象，归根结底是需要"通用"的缘故。比如网上资源是一个通用需求，每个家庭都相当于容器的一个结点，那么通信设备比如电话线等就相当于迭代器。千家万户通过上网就可以查询所需要的信息。再比如生活中用到的自来水是通用需求，每个家庭仍旧相当于一个结点，那么水管线等就相当于迭代器。

既然生活中的通用需求促进了各种通信事业等的发展，那么对于软件中的通用算法而言，也就一定能促进迭代器的不断进步。因此，希望同学们多观察生活现象，因为软件的许多设计思想就在我们的生活之中。如果软件中的每个设计思想都能找到生活中的实例，那么设计软件就一定能使你感到非常快乐，就不那么枯燥了。

3.4 STL 迭代器

3.1 节～3.3 节中通过自定义迭代器理解了 STL 迭代器的一些基本思想，只不过功能较简单，迭代器类还有许多待完善的地方。按照功能划分，STL 迭代器共分为 5 大类型。

（1）输入迭代器(Input Iterator)。按顺序只读一次。完成的功能有：能进行构造和默认构造，能被复制或赋值，能进行相等性比较，能进行逐步向前移动，能进行读取值。输入迭代器重载主要操作符如表 3.1 所示。

表 3.1 输入迭代器重载主要操作符

操 作 符	说 明	操 作 符	说 明
operate *	访问迭代元素值	operate==	迭代元素相等比较
operate++()	前置迭代指针++	operate!=	迭代元素不等比较
operate++(int)	后置迭代指针++		

STL 提供的主要输入迭代器是 istream_iterator，支持表 3.1 中的所有操作。值得注意的是它的构造函数，有两种形式：

- istream_iterator()：默认的构造器，创建了一个流结束的迭代器。
- istream_iterator(istream &)：参数是输入流。含义是从输入流中读数据，当遇到流结束符时停止。

【例 3.2】 利用 istream_iterator 迭代器迭代标准输入流。

```
//文件名：e3_2.cpp
#include<iostream>
#include<iterator>
using namespace std;
int main(int argc, char * argv[])
```

```
    {
        cout<<"请输入数据(如 111 222 333,):";
        istream_iterator<int>a(cin);
                                         //建立键盘输入流,并用 istream_iterator 枚举整型数据
        istream_iterator<int>b;          //建立输入流结束迭代器
        while(1)
        {
            cout<< * a<<endl;            //输出整型数据→调用 operator * ()
            a++;                         //迭代器指针指向下一个元素→调用 operator++(int)
            if(a==b)                     //如果当前迭代器等于结束迭代器,则→调用 operator==
            {                            //退出 while 循环
                break;
            }
        }
        return 0;
    }
```

程序执行结果为:

请输入数据(如111 222 333,): 111 222 333,
   111
   222
   333

程序分析如下所示。

① 先从键盘(标准输入)输入相应数据,按 Enter 键后,调用 istream_iterator 迭代器进行枚举数据,迭代器指针指向第一个符合要求的数据。

② 利用循环输出枚举数据,若迭代器指针已到达数据流尾部,则退出循环。那么如何判断到达数据流尾部呢? 从程序中可看出如果迭代指针 a 等于 b 即可,而 b 即是调用无参数的 istream_iterator 构造函数获得的。

注意:本例中输入的是"111 222 333,",后面跟一个逗号,若无逗号则 istream_iterator<int> a(cin)不会执行完,这是由于迭代类型是整型数,只有在流中输入非整型数,迭代器迭代到此位置时认为它不是整型数,这时才停止工作。本例最后输入了",",当然只要是其他的非整型数都是可以的。

(2) 输出迭代器(Output Iterator)。只写一次。完成的功能有:能进行构造或默认构造,能被复制或赋值,能进行相等性比较,能进行逐步前向移动,能进行写入值(* p＝x,但不能读出)。输出迭代器重载主要操作符如表 3.2 所示。

表 3.2 输出迭代器重载主要操作符

操 作 符	说 明	操 作 符	说 明
operate *	分配迭代元素值空间	operate＋＋()	前置迭代指针＋＋
operate＝	写入元素值	operate＋＋(int)	后置迭代指针＋＋

STL 提供的主要输出迭代器是 ostream_iterator,支持表 3.2 中的所有操作。值得注意的是它的构造函数,有两种形式:

- ostream_iterator(ostream& out):创建了流输出迭代器,用来迭代 out 输出流。
- ostream_iterator(ostream& out, const char * delim):创建了流输出迭代器,用来向 out 输出流输出数据,输出的数据之间用 delim 字符串分隔。即每向 out 输出流输出一个数据后,就向 out 输出流输出一个分隔符 delim。

【例 3.3】 利用 ostream_iterator 向屏幕输出数据。

```
//文件名:e3_3.cpp
#include<iostream>
#include<iterator>
using namespace std;

int main(int argc, char * argv[])
{
    cout<<"输出迭代器演示结果为:";
    ostream_iterator<int>myout(cout, "\t");          //创建标准输出迭代器
    * myout=1;
    myout++;
    * myout=2;
    myout++;
    * myout=3;
    return 0;
}
```

执行结果为:

输出迭代器演示结果为: 1   2   3

(3) 前向迭代器(Forward Iterator)。使用输入迭代器和输出迭代器可以基本满足算法和容器的要求,但还是有一些算法需要同时具备两者的功能,例如 replace 算法就是一例。前向迭代器就是为满足这些需求而定义的。先看看 replace 算法(用新值代替旧值)的需求。

```
template<class ForwardIterator, class T>
void replace (ForwardIterator first, ForwardIterator last,
    const T&old_value, const T& new_value)
{
    for (;first!=last;++fist)
        if ( * first==old_value)
            * first=new_value;
}
```

这里的 first、last 就是 Forward Iterator 的模型,其需求包括 first!=last、++first、

*first==old_value、*first=new_value。这样,Forward Iterator 的需求概括为:能进行构造或默认构造,能被复制或赋值,能进行相等性比较(对应满足 first!=last 的需求),能进行逐步前向移动(对应满足++first 的需求),能进行读取值(x=*p,但不能改写,对应满足*first 的需求)。前向迭代器包含了输入和输出迭代器两者的所有功能,加上还可以多次解析一个迭代器指定的位置,因此可以对一个值进行多次读写。顾名思义,前向迭代器只能向前移动。但是 STL 本身并没有专为前向迭代器预定义的迭代器。

(4) 双向迭代器(Bidirectional Iterator)。具有前向迭代器的全部功能,另外它还可以利用自减操作符 operate－－向后一次移动一个位置。例如双向链表容器中需要的就是双向迭代器。

(5) 随机访问迭代器(Random Access Iterator)。具有双向迭代器的所有功能,再加上一个指针所有的功能。包括使用操作符 operator[]进行索引,加某个整数值到一个指针就可以向前或向后移动若干个位置,或者使用比较运算符在迭代器之间进行比较。

综观 5 种 Iterator,发现从前到后需求越来越多,也就是所谓的细化。这样,在一切情况下都可以使用需求最细的 Random Access Iterator,确实可以,但不好,因为过多的需求自然会降低它的效率,实际编程时应该选择正好合适的 Iterator 以期得到最高的效率。

第 4 章 输入输出流

在 C++ 的标准模板库中提供了一组模板类来支持面向对象的数据输入输出功能,如基本的输入输出流 istream 类/ostream 类/iostream 类、文件输入输出流 ifstream 类/ofstream 类/fstream 类、字符串输入输出流 stringstream 类/istringstream 类/ostringstream 类等。C++ I/O 还可以对对象进行输入输出操作,这些都是 C 函数所不具备的。这些流类都位于 std 名字空间内,一般都是用 stream 作后缀名。它们在数据流的处理上具有灵活、完善和强大的功能。

4.1 标准输入输出流

标准输入输出流即标准输入流 cin、标准输出流 cout,前者指键盘,后者指显示器。

4.1.1 插入符与提取符

在输入输出流类库中,重载了两种运算符以简化输入输出流的使用:运算符<<常用做输入输出流的插入符,表明"输出到",例如 cout<<"Hello"是把字符串"Hello"输出到屏幕上;运算符>>常用做提取符,表明"赋值给",例如 cin>>i 是把键盘输入的信息赋值给 i。

【例 4.1】 标准输入输出流给不同变量赋值。

```
//文件名:e4_1.cpp
#include<iostream>
using namespace std;
int main() {
    int i;
    float f;
    char str[20];
    cin>>i;
    cin>>f;
    cin>>str;
```

```
    cout<<"i="<<i<<endl;
    cout<<"f="<<f<<endl;
    cout<<"str="<<str<<endl;

    return 0;
}
```

本程序的功能是从键盘依次输入整型数、浮点数、字符串,并通过 cin 解析键盘输入流,分别赋值给整型数 i、浮点数 f、字符串 str,最后通过 cout 输出流再把各个值显示在屏幕上。

当然,从示例中也能总结出以下知识点:

(1) 插入符<<、提取符>>均可连续插入,连续赋值。对于 cin 而言,各数值一般默认是以空格为分界符的。示例中三个独立行 cin>>i; cin>>f; cin>>str;可以合并成一行 cin>>i>>f>>str;。对于 cout 而言,endl 表示回车换行。

(2) 可以通过插入符<<输入不同类型的数据,可以通过提取符>>给不同类型的数据赋值。

【例 4.2】 标准输入给不同类型变量赋值的不足示例。

```
//文件名:e4_2.cpp
#include<iostream>
using namespace std;
int main() {
int i;
char str[20];

cout<<"请输入一个整型数及一个字符串:";
cin>>i>>str;
cout<<"i="<<i<<endl;
cout<<"str="<<str<<endl;
return 0;
}
```

若在命令行上输入 1 how are you?,按 Enter 键后执行结果为:

请输入一个整型数及一个字符串:1 how are you?
i=1
str=how

可以看出 str 等于 how,并非"how are you?"。这是由于命令行各参数默认是以空格为界定符的,1 正常赋值给 i 了。由于 how 后是空格,因此解析串到此位置停止了,所以 str 仅等于 how。

实际上在交互过程中,经常需要一次输入一行字符序列,当这些字符安全地存储到缓冲区后再进行扫描和转换工作。在例 4.2 中,并不能把输入的串完全地放到内存中,下面讲到的 get 系列函数很好地解决了这个问题。

4.1.2 get 系列函数

常用的 get 系列函数有三个。
(1) int get();
返回输入流一个字符的 ASCII 值。
(2) istream& get(unsigned char * pszBuf, int nBufLen, char delim='\n')
- pszBuf:指向字符缓冲区的指针,用于保存结果。
- nBufLen:缓冲区长度。
- delim:结束字符,根据结束字符判断何时停止读入操作。

(3) istream& getline(unsigned char * pszBuf, int nBufLen, char delim='\n');
参数解释同 get。

第二个 get 函数及 getline 函数都是读一行记录,那么它们有什么区别呢? 细微而重要的区别在于:当遇到输入流中的界定符(delim,即结束字符)时,get()停止执行,但是并不从输入流中提取界定符,直接在字符串缓冲区尾部加结束标志\0;函数 getline()则相反,它将从输入流中提取界定符,但不会把它存储到结果缓冲区中。

【例 4.3】 get、getline 函数简单示例。

```
//文件名:e4_3.cpp
#include<iostream>
using namespace std;
int main() {
    char szBuf[60];                            //定义输入字符串接收缓冲区
    cout<<"请输入一个字符串:";
    int n=cin.get();                           //先读一个字符
    cin.getline(szBuf, 60);                    //接着读一行字符,遇到结束符"\0"停止
    cout<<n<<endl;
    cout<<"The received string is: "<<szBuf<<endl;
    return 0;
}
```

执行结果为:

请输入一个字符串:How are you?
72
The received string is:ow are you?

分析如下:首先键盘输入串"How are you?",当执行到 cin.get()后,读取输入流第一个字符的 ASCII 码,由于"H"的 ASCII 值是 72,因此当执行 cout<<n<<endl 时,屏幕输

出 72；之后利用 cin.getline(szBuf，60)继续读剩余的输入流字符串，由于 getLine()函数第三个参数默认结束符是\n，因此 szBuf 的值是"ow are you?"。

4.1.3 处理流错误

例如应该输入一个整型数据，可是却误输了字符串，如何来判断此类错误呢？C++对每次输入输出操作的结果都记录了其状态信息，编程者通过这些状态信息即可判断此次输入输出操作正确与否。

获取状态信息的函数如下：
int rdstate();；无参数，返回值即是状态信息特征值。
使用下面函数来检测相应输入输出状态：
- bool good();；若返回值 true，一切正常，没有错误发生。
- bool eof();；若返回值 true，表明已到达流的末尾。
- bool fail();；若返回值 true，表明 I/O 操作失败，主要原因是非法数据（例如读取数字时遇到字母）。但流可以继续使用。
- bool bad();；发生了（或许是物理上的）致命性错误，流将不能继续使用。

【例 4.4】 判断是否输入整型数。

```
//文件名：e4_4.cpp
#include<iostream>
using namespace std;
int main() {
    int a;
    cout<<"请输入一个数据：";
    cin>>a;
    cout<<"状态值为: " <<cin.rdstate()<<endl;
    if(cin.good())
    {
        cout<<"输入数据的类型正确,无错误!"<<endl;
    }
    if(cin.fail())
    {
        cout<<"输入数据类型错误,非致命错误,可清除输入缓冲区挽回!"<<endl;
    }
    return 0;
}
```

当输入不同的数据（例如 10 或 abcde）后，程序显示结果为：

第一种情况   第二种情况
请输入一个数据：10   请输入一个数据：abcde

状态值为：0	状态值为：2
输入数据的类型正确，无错误！	输入数据类型错误，非致命错误，可清除输入缓冲区挽回！

通常当发现输入有错又需要改正的时候，使用 clear()更改标记改正确后，同时需要使用 get()成员函数清除输入缓冲区，以达到重复输入的目的。

【例 4.5】 确保输入一个整型数给变量 a。

```cpp
//文件名：e4_5.cpp
#include <iostream>
using namespace std;
int main() {
    int a;
    while(1)
    {
        cin>>a;                              //赋值给整型数 a
        if(cin.fail())                       //如果输入数据非法，则：
        {
            cout<<"输入有错！请重新输入"<<endl;
            cin.clear();                     //清空状态标识位
            cin.get();                       //清空流缓冲区
        }
        else                                 //如果输入数据合法，则：
        {
            cout<<a<<endl;                   //直接在屏幕上输出数据
            break;                           //跳出 while 循环
        }
    }
    return 0;
}
```

其中，仅当输入数据错误时给出提示信息。在此基础上，通过 while 循环给出了必须输入正确数据类型的方法，程序执行结果如下所示。

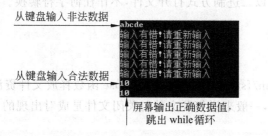

那么为什么屏幕上显示 5 次提示信息？这是由于输入的 abcde 串长是 5，而读输入流函数 cin.get()一次只能读入一个字符。因此，while 循环执行了 5 次，显示了 5 次提示信息，这时输入流已至末尾，当再次执行 while 循环时，遇到 cin>>a，则必须从键盘再次输入数据。当然，可以用 getline 函数替换 get 函数，一次清空缓冲区，显示一次提示信息，也是没

有问题的。

4.2 文件输入输出流

C++对文件的读写使用的是 ifstream、ofstream 和 fstream 流类,同样是依照"打开→读写→关闭"原语进行操作。文件读写所用到的很多常数都在基类 ios 中被定义出来。ifstream 类只用于读取文件数据,ofstream 类只用于写入数据到文件,fstream 类则可用于读取和写入文件数据。

4.2.1 文件打开

ifstream/ofstream/fstream 流类都使用构造函数或 open 函数打开文件,原型如下:
- ifstream(const char * szName, int nMode=ios::in, int nProt=filebuf::openprot)
- ofstream(const char * szName, int nMode=ios::out, int nProt=filebuf::openprot)
- fstream(const char * szName, int nMode, int nProt=filebuf::openprot)
- void ifstream::open(const char * filename, int openmode=ios::in)
- void ofstream::open(const char * filename, int openmode=ios::out|ios::trunc)
- void fstream::open(const char * filename, int openmode=ios::in|ios::out)

均有两个参数,第一个是文件名,第二个是打开方式,打开方式如下所述:
- ios::in:以读取方式打开文件。
- ios::out:以写入方式打开文件。
- ios::app:每次写入数据时,先将文件指针移到文件尾,以追加数据到尾部。
- ios::ate:仅初始时将文件指针移到文件尾,仍可在任意位置写入数据。
- ios::trunc:写入数据前,先删除文件原有内容(清空文件),当文件不存在时会创建文件。
- ios::binary:以二进制方式打开文件,不作任何字符转换。

4.2.2 文件关闭

ifstream/ofstream/fstream 流类都通过 close 函数释放文件资源。
close();无参数,一般来说打开文件与关闭文件是成对出现的。

4.2.3 文件读写

1. 读写文本文件

文本文件是最常见的操作对象,关键是要解决如何按行读、如何按行写的问题。

【例 4.6】 读文本文件并显示在屏幕上。

```cpp
//文件名：e4_6.cpp
#include<fstream>
#include<iostream>
using namespace std;

int main(){
    char szBuf[80];
    ifstream in("a.txt");           //通过构造函数创建文件读入流
    if(!in) return 0;               //若文件不存在,返回

    while(in.getline(szBuf, 80))    //通过getline函数按行读取内容
    {
        cout<<szBuf<<endl;          //将读入缓冲区内容输出到屏幕上,判断是否正确
    }
    in.close();
    return 0;
}
```

可以看出按行读主要是通过 getline 函数来完成的。

【例 4.7】 写文本文件：把学生成绩信息保存至文件。

```cpp
//文件名：e4_7.cpp
#include<fstream>
#include<iostream>
using namespace std;
struct STUDENT                      //学生成绩结构体
{
    char strName[20];               //姓名
    int nGrade;                     //成绩
};
int main() {
    ofstream out;
    out.open("d:\\a.txt");          //打开或创建 a.txt 文本文件
    STUDENT st1={"张三", 90};       //用结构体产生两名学生的成绩信息
    STUDENT st2={"李四", 80};
    out<<st1.strName<<"\t"<<st1.nGrade<<endl;   //把成绩信息存到文本文件
    out<<st2.strName<<"\t"<<st2.nGrade<<endl;

    out.close();
    return 0;
}
```

可以看出,通过<<完成写文本文件是一个较好的方法,这是因为<<操作符可把不同类型的数据直接输出。结构体中 strName 是字符串,nGrade 是整型数,但均可直接输出到文件上。此外,strName 与 nGrade 之间采用\t 输出是为了数据对齐,而 endl 保证输出\n 回车符,保证了文本文件的换行,是必需的。

2. 读写二进制文件

主要是通过 read、write 函数完成读写二进制文件的功能,原型如下所示:

- ostream& write(const char * , int nSize)
- istream& read(char * , int nSize)

第一个参数表明读写缓冲区的头指针,第二个参数表明读写缓冲区的大小。

【例 4.8】 写二进制文件:把学生成绩信息保存至文件。

```
//文件名:e4_8.cpp
#include<fstream>
#include<iostream>
using namespace std;
struct STUDENT                              //学生成绩结构体
{
    char strName[20];                       //姓名
    int nGrade;                             //成绩
};
int main() {
    ofstream out;
    out.open("d:\\a.txt");                  //打开或创建 a.txt 文本文件
    STUDENT st1={"张三", 90};               //用结构体产生两名学生的成绩信息
    STUDENT st2={"李四", 80};
    out.write((const char * )&st1, sizeof(STUDENT));    //把成绩信息存到二进制文件
    out.write((const char * )&st2, sizeof(STUDENT));

    out.close();
    return 0;
}
```

由于 STUDENT 结构体的大小是 24,因此保存两名学生信息的二进制码文件大小是 48,与结果是相符的。用 type 命令显示该文件,可以看出是乱码。

【例 4.9】 读二进制文件,并把结果显示在屏幕上。

```
//文件名:e4_9.cpp
#include<fstream>
#include<iostream>
using namespace std;
struct STUDENT
```

```cpp
    {
        char strName[20];                            //姓名
        int nGrade;                                  //成绩
    };
    int main() {
        ifstream in;
        in.open("d:\\a.txt");
        if(!in) return 0;

        STUDENT st1;
        STUDENT st2;
        in.read((char*)&st1, sizeof(STUDENT));
        in.read((char*)&st2, sizeof(STUDENT));

        cout<<st1.strName<<"\t"<<st1.nGrade<<endl;
        cout<<st2.strName<<"\t"<<st2.nGrade<<endl;

        in.close();
        return 0;
    }
```

3. 输入输出流缓冲

C++标准库封装了一个缓冲区类 streambuf，以供输入输出流对象使用。每个标准 C++ 输入输出流对象都包含一个指向 streambuf 的指针，用户可以通过调用 rdbuf() 成员函数获得该指针，从而直接访问底层 streambuf 对象；可以直接对底层缓冲区进行数据读写，从而跳过上层的格式化输入输出操作。但由于类似的功能均可由上层缓冲区类实现，因此就不再加以论述了。streambuf 最精彩的部分在于它重载了 operator<< 及 operator>>。对 operator<< 来说，它以 streambuf 指针为参数，实现把 streambuf 对象中的所有字符输出到输出流中；对 operator>> 来说，可把输入流对象中的所有字符输入到 streambuf 对象中。

下面是一个简单的示例程序。

【例 4.10】 打开一个文件并把文件中的内容送到标准输出流中。

```cpp
//文件名：e4_10.cpp
#include<fstream>
#include<iostream>
using namespace std;
int main() {
    ifstream fin("d:\\my.txt");
    if(fin!=NULL)                                   //判断文件是否存在
```

```
        cout<<fin.rdbuf();                          //把文件所有内容输出到屏幕上
    }
    fin.close();
}
```

同样是完成文件输出,此方法是非常简洁的。当然,把 cout<<fin.rdbuf()左侧的标准输出改为文件输出流等,则可实现文件的复制功能等。

4. 定位输入输出流

流的位置标识有三个。
- ios::beg:流的开始位置。
- ios::cur:流的当前位置。
- ios::end:流的结束位置。

定位函数主要有两个:

① istream& seekg(long relativepos, ios::seek_dir dir)

针对输入流。第一个参数是要移动的字符数目,可正可负;第二个参数是移动方向,是 ios::beg、ios::cur、ios::end 中的一个值;含义是字符指针相对于移动方向向前或向后移动了多少个字符。

② ostream& seekp(long relativepos, ios::seek_dir dir)

针对输出流。参数含义同 seekg。

【例 4.11】 下面示例先写文件,然后再把文件内容读出显示在屏幕上。

```
//文件名:e4_11.cpp
#include<fstream>
#include<iostream>
using namespace std;
int main() {
    fstream    in_out;
    in_out.open("d:\\my.txt",ios::in|ios::out|ios::trunc);
    in_out.write("Hello", 5);

    in_out.seekg(0, ios::beg);                       //读指针移到文件头
    cout<<in_out.rdbuf();
    in_out.close();
    return 0;
}
```

对该程序分析如下。

(1) 采用了 fstream 输入输出流类,既可读又可写。

(2) 注意 open 打开标志 ios::in|ios::out|ios::trunc,特别是 ios::trunc,它保证了若在指定位置没有该文件,则自动创建该文件。

(3) 文件打开或创建成功后,文件指针均指向文件头,当写完字符串"Hello"后,文件指

针已经偏移了,若想完全显示全部文件内容,必须把指针移到文件头。文中用到了 seekg 函数,其实用 seekp 函数也是等效的,这是因为 fstream 是输入输出流。若是单独的输入流,则只能用 seekg 函数;若是单独的输出流,则只能用 seekp 函数。

4.3 字符串输入输出流

字符串输入输出流类直接对内存而不是对文件和标准输出进行操作,它使用与 cin 及 cout 相同的读取和格式化函数来操纵内存中的数据,所有字符串流类的声明都包含在标准头文件<sstream>中。标准库定义了三种类型的字符串流。

- istringstream:字符串输入流,提供读 string 功能。
- ostringstream:字符串输出流,提供写 string 功能。
- stringstream:字符串输入输出流,提供读写 string 功能。

利用字符串输入输出流,可以方便地把多种基本数据类型组合成字符串,也可以反解字符串给各种变量赋值。

【例 4.12】 反解字符串给各变量赋值。

```
//文件名:e4_12.cpp
#include<iostream>
#include<sstream>
using namespace std;

int main()
{
    int n;
    float f;
    string strHello;

    string strText="1 3.14 hello";
    istringstream s(strText);
    s>>n;
    s>>f;
    s>>strHello;

    cout<<"n="<<n<<endl;
    cout<<"f="<<f<<endl;
    cout<<"strHello="<<strHello<<endl;
    return 0;
}
```

可知,通过字符串输入流 istringstream 解析字符串"1 3.14 hello",依次赋给整型、浮点型、字符串变量,这种方法是很简便的。

【**例4.13**】 合并不同类型的数据到字符串。

```
//文件名:e4_13.cpp
#include<iostream>
#include<sstream>
using namespace std;

int main()
{
    cout<<"type an int,a float and a string: ";
    int i;
    float f;
    string stuff;
    cin>>i>>f;
    getline(cin, stuff);
    ostringstream    os;
    os<<"integer="<<i<<endl;
    os<<"float="<<f<<endl;
    os<<"string="<<stuff<<endl;

    string result=os.str();
    cout<<result<<endl;
    return 0;
}
```

执行结果为:

```
type an int, a float, a string: 1 3.14 how are you
integer=1
float=3.14
string=how are you
```

4.4 综合示例

尽管cin、cout可以自动对基本数据类型或字符串string等进行输入输出操作,但是在某些情况下若想对类对象直接进行输入输出操作是不行的,必须重载operator<<及operator>>。示例如下。

【**例4.14**】 从键盘中输入学生属性值,并在屏幕上重新显示。

```
//文件名:e4_14.cpp
#include<iostream>
```

```cpp
#include<string>
using namespace std;

class Student
{
public:
    string strName;                                         //姓名
    string strSex;                                          //性别
    int    nAge;                                            //年龄
};
istream& operator>>(istream& is, Student& s)
{
    is>>s.strName>>s.strSex>>s.nAge;
    return is;
}
ostream& operator<<(ostream& os, Student& s)                //Student是普通对象
{
    os<<s.strName<<"\t"<<s.strSex<<"\t"<<s.nAge<<"\n";
    return os;
}
ostream& operator<<(ostream& os, const Student& s)          //Student是常对象
{
    cout<<"const Student 输出是:"<<endl;
    os<<s.strName<<"\t"<<s.strSex<<"\t"<<s.nAge<<"\n";
    return os;
}
void f(const Student& s)
{
    cout<<s;
}
int main(int argc, char * argv[])
{
    Student s;
    cout<<"请输入数据(姓名 性别 年龄):";
    cin >>s;
    cout<<"输入的数据是：";
    cout<<s;
    f(s);
    return 0;
}
```

程序执行结果为：

请输入数据(姓名 性别 年龄):张三　男　20

输入的数据是:张三 男 20
const Student 输出是:张三 男 20

对该程序着重理解主程序中 cin>>s、cout<<s 的用法,它们均是对对象的操作。当 cin >> s 时,其实真正调用的是重载的 istream& operator>>(istream& is, Student& s),如何输入是由该函数体来完成的;同理,当 cout<<s 时,真正调用的是重载的 ostream& operator<<(ostream& os, Student& s),如何输出由该函数体来完成。当 Student 是常对象时,例如函数 void f(const Student& s)中 cout<<s 调用的是 ostream& operator<<(ostream& os, const Student& s)。那么为什么对基本数据类型就不必重载相应函数呢?那是因为 STL 中默认实现了基本数据类型的输入输出函数,就不必再重载了。

【例 4.15】 编写一个程序,从文本文件中读入每个学生的各科成绩,并在屏幕上显示学生各科成绩及总成绩。假设文本文件格式如表 4.1 所示。

表 4.1 文本文件格式示例

张三	60	70	80
李四	70	80	90
⋮			

方法 1:

```cpp
//文件名:e4_15_1.cpp
#include<iostream>
#include<fstream>
#include<sstream>
#include<string>

using namespace std;
int main(int argc, char * argv[])
{
    ifstream in("d:\\data.txt");            //文件以读方式打开

    string strText;
    string strName;                         //学生姓名
    int nYuwen;                             //语文成绩
    int nMath;                              //数学成绩
    int nForeign;                           //外语成绩
    int nTotal;                             //总成绩
    while(!in.eof())
    {
        getline(in, strText);               //读每行文本数据保存至字符串
        istringstream inf(strText);         //把该文本串封装成字符串输入流对象
```

```
        inf>>strName>>nYuwen>>nMath>>nForeign;        //通过字符串输入流对象
                                                      //给姓名及各成绩赋值
        nTotal=nYuwen+nMath+nForeign;
        cout<<strName<<"\t"<<nYuwen<<"\t"<<nMath<<"\t"<<nForeign <<"\t"<<
        nTotal<<endl;
    }
    in.close();
    return 0;
}
```

思路是：首先用 open 函数以读方式打开该文件；然后用 getline 函数一行行读，把每行内容存至字符串，并把该静态的字符串封装成动态的字符串输入流对象；再以流的方式给学生姓名及各科成绩赋值，算出总成绩，输出至屏幕。

方法 2：

```
//文件名：e4_15_2.cpp
#include<iostream>
#include<fstream>
#include<sstream>
#include<string>
using namespace std;

class Student
{
public:
    string strName;                                  //姓名
    int nYuwen;                                      //语文
    int nMath;                                       //数学
    int nForeign;                                    //外语
};

istream& operator>> (istream& is, Student& s)
{
    is>>s.strName>>s.nYuwen>>s.nMath>>s.nForeign;
    return is;
}

ostream& operator<< (ostream& os, Student& s)
{
    int nTotal=s.nYuwen+s.nMath+s.nForeign;
    os<<s.strName<<"\t"<<s.nYuwen<<"\t"<<s.nMath<<"\t"<<s.nForeign<<"\t"<<
    nTotal<<"\n";
    return os;
```

```
    }
    int main(int argc, char * argv[])
    {
        ifstream in("d:\\data.txt");
        Student s;
        while(!in.eof())
        {
            in>>s;
            cout<<s;
        }
        in.close();
        return 0;
    }
```

该方法同方法1在设计思想上是不同的。方法1仅仅完成了题目所述功能,从代码上看不出每行数据代表一个学生的成绩信息,只是一个个孤立的数据;但方法2就不同了,关键在于它定义了学生类Student,从主程序中in>>s,cout<<s也可明确看出要给学生对象赋值,要把学生对象信息输出到屏幕。

也许有的同学认为方法1代码更简洁。但是从面向对象角度来看,方法2更容易扩展和维护。对方法1来说,如果将来的需求分析发生变化,也许改动得更多。

第 5 章 字 符 串

字符串是程序设计中最复杂的编程内容之一。STL string 类提供了强大的功能,使得许多烦琐的编程内容用简单的语句就可完成。string 字符串类减少了 C 语言编程中三种最常见且最具破坏性的错误:超越数组边界;通过未被初始化或被赋以错误值的指针来访问数组元素;以及在释放了某一数组原先所分配的存储单元后仍保留的"悬挂"指针。

5.1 字符串创建及初始化

5.1.1 基本创建方式

【例 5.1】 字符串基本创建示例。

```
//文件名:e5_1.cpp
#include<string>
using namespace std;
int main(int argc, char * argv[])
{
    string s1;
    string s2("How are you?");
    string s3(s2);
    string s4(s2, 0, 3);
    string s5="Fine";
    string s6=s2+"Fine";
    return 0;
}
```

对象 s1~s4 均是通过构造函数的方式创建新字符串对象。主要理解点如下。

(1) s1 string 对象被创建,但不包含初始值。C 语言中的 char 型数组在初始化前都包含随机的无意义的位模式,而与此不同,s1 确实包含了有意义的信息。这个 string 对象被初始化成包含"没有字符",通过类的成员函数能够正确地报告其长度为 0 并且没有数据元素。

（2）创建 s4 对象，构造函数中有三个参数。第一个参数类型是 string 类型，第二、三个参数是整型，分别表示偏移量及计数量。该构造函数的含义是源串 s2，从偏移量 0 的字符开始连续取三个字符，构成新的字符串对象，因此 s4＝"How"。

（3）对象 s5，s6 通过赋值的方式产生新的对象。s5 是单一赋值，可直接把 char 型数组赋给 s5；s6 表明可将不同的初始化数据源结合在一起，本例即是把一个 string 对象 s2 与一个 char 型数组结合，构成一个新的对象 s6。但要注意：写成 string s6＝s2＋"Fine"可编译通过，而写成 string s6＝"Fine"＋s2 不能编译通过。从中可以看出：string 对象赋值，等号右边第一项必须是 string 类型，不能是 char 型数组。

5.1.2 迭代器创建方式

由于可将 string 看做字符的容器对象，因此可以给 string 类的构造函数传递两个迭代器，将它们之间的数据复制到新的 string 对象中。

【例 5.2】 字符串迭代器创建方式。

```
//文件名：e5_2.cpp
#include<string>
using namespace std;
int main(int argc, char * argv[])
{
    string s1="How are you?";
    string s2(s1.begin(), s1.end());
    string s3(s1.begin()+4,s1.begin()+7);
    return 0;
}
```

由于 s1.begin()，s1.end()迭代器覆盖了 s1 的全部内容，因此 s2＝"How are you?"；s1.begin()＋4，s1.begin()＋7 仅代表了 s1 部分内容"are"，因此 s3＝"are"。

5.2 字符串操作

5.2.1 插入操作

字符串一般包括首字符前、尾字符后、任意位置插入等几种情况。

【例 5.3】 字符串插入操作。

```
//文件名：e5_3.cpp
#include<string>
#include<iostream>
```

```
using namespace std;
int main(int argc, char * argv[])
{
    string s="do";
    cout<<"Inition size is:"<<s.size()<<endl;
    s.insert(0, "How");
    s.append("you");
    s=s+"do?";

    cout<<"final size is:"<<s.size()<<endl;
    cout<<s;
    return 0;
}
```

执行结果如下：

```
Inition size is:2
final size is:14
How do you do?
```

通过该示例，可得出主要知识点如下。

(1) insert 函数，第一个参数表明插入源串的位置，第二个参数表明要插入的字符串，因此利用该函数可实现串首、串尾及任意位置处的字符串插入功能。

(2) append 函数，仅有一个输入参数，在源字符串尾部追加该字符串。

(3) 利用+实现字符串的连接，从而创建新的字符串。

(4) size 函数，无输入参数，通过例子可知：它表明字符串长度值，即有多少个字符，初始的时候值是 2，当完成增加后，值就变成 14 了。这说明字符串 string 类本身可根据需要自动调节串所在的内存空间大小，编程者无须参与，这一点是 C 语言中 char 型数组无法比拟的。

5.2.2 替换操作

常用的是 replace 函数，有三个输入参数：第一个用于指示从字符串的什么位置开始改写；第二个用于指示从原字符串中删除多少个字符，第三个是替换字符串的值。

【例 5.4】 字符串替换操作。

```
//文件名：e5_4.cpp
#include<string>
#include<iostream>
using namespace std;
int main(int argc, char * argv[])
{
```

```
string s="what's your name?";
cout <<"替换前:"<<s<<endl;
s.replace(7, 4, "her");
cout <<"替换后:"<<s<<endl;
return 0;
}
```

执行结果为:

替换前:what's your name?
替换后:what's her name?

5.3 字符串查询

主要掌握的知识点如下。

- string::npos:这是 string 类中的一个成员变量,一般应用在判断系统查询函数的返回值上,若等于该值,表明没有符合查询条件的结果值。
- find 函数:在一个字符串中查找指定的单个字符或字符组。如果找到,就返回首次匹配的开始位置;如果没有找到匹配的内容,则返回 string::npos。一般有两个输入参数,一个是待查询的字符串,一个是查询的起始位置,默认起始位置为 0。
- find_first_of 函数:在一个字符串中进行查找,返回值是第一个与指定字符串中任何字符匹配的字符位置;如果没有找到匹配的内容,则返回 string::npos。一般有两个输入参数,一个是待查询的字符串,一个是查询的起始位置,默认起始位置为 0。
- find_last_of 函数:在一个字符串中进行查找,返回最后一个与指定字符串中任何字符匹配的字符位置;如果没有找到匹配的内容,则返回 string::npos。一般有两个输入参数,一个是待查询的字符串,一个是查询的起始位置,默认起始位置为 0。
- find_first_not_of 函数:在一个字符串中进行查找,返回第一个与指定字符串中任何字符都不匹配的元素位置;如果没有找到匹配的内容,则返回 string::npos。一般有两个输入参数,一个是待查询的字符串,一个是查询的起始位置,默认起始位置为 0。
- find_last_not_of 函数:在一个字符串中进行查找,返回下标值最大的与指定字符串中任何字符都不匹配的元素位置;如果没有找到匹配的内容,则返回 string::npos。一般有两个输入参数,一个是待查询的字符串,一个是查询的起始位置,默认起始位置为 0。
- rfind 函数:对一个串从尾至头查找指定的单个字符或字符组,如果找到,就返回首次匹配的开始位置;如果没有查找到匹配的内容,则返回 string::npos。一般有两个输入参数,一个是待查询的字符串,一个是查询的起始位置,默认起始位置为串尾部。

【例 5.5】 字符串查询函数基本用法。

```cpp
//文件名: e5_5.cpp
#include<string>
#include<iostream>
using namespace std;

int main(int argc, char * argv[])
{
    string s="what't your name?my name is TOM. How do you do?Fine, thanks. ";
    int n=s.find("your");
    cout<<"the first your pos:"<<n<<endl;
    n=s.find("you", 15);
    cout<<"the first your pos begin from 15:"<<n<<endl;
    n=s.find_first_of("abcde");
    cout<<"find pos when character within abcde:"<<n<<endl;
    n=s.find_first_of("abcde", 3);
    cout<<"find pos begin from 2 when character within abcde:"<<n<<endl;
    return 0;
}
```

执行结果如下：

```
the first your pos: 7
the first your pos begin from 15:41
find pos when character within abcde:2
find pos begin from 2 when character within abcde:13
```

(1) find("your")：从源串起始位置为 0（默认值）处查找有"your"字符串位置，所以结果为 7。

(2) find("you", 15)：从源串起始位置为 15 处查找有"you"字符串位置，所以结果为 41。

(3) find_first_of("abcde")：从源串起始位置为 0（默认值）处依次查找每个字符，如果它在输入的字符串参数"abcde"中，则返回该字符的位置。由于源串头两个字符"w"、"h"不在查找的串中，而第三个字符"a"在查找的目的串中，又由于字符串是以 0 为基点的，所以结果值为 2。

(4) find_first_of("abcde",3)：从源串起始位置为 3 处依次查找每个字符，如果它在输入的字符串参数"abcde"中，则返回该字符的位置。分析同 find_first_of("abcde")，结果为 13，对应的源串字符为"a"。

本例是以 find、find_first_of 函数为例来写的，同理可写出近似的 find_last_of()、find_first_not_of()、find_last_not_of()、rfind()代码，同学们可自行完成。

5.4 在字符串中删除字符

在字符串中删除字符主要用 erase 函数,有两个迭代器输入参数,之间表示的字符将被删除掉。

【例 5.6】 字符串删除函数示例。

```
//文件名：e5_6.cpp
#include<string>
#include<iostream>
using namespace std;
int main(int argc, char * argv[])
{
    string s1="How are you?";
    s1.erase(s1.begin(), s1.begin()+3);
    cout<<"after erase to s1 is:"<<s1<<endl;

    string s2="Fine, thanks";
    s2.erase(s2.begin(), s2.end());
    cout<<"after erase to s2 is:"<<s2<<endl;
    return 0;
}
```

执行结果为：

after erase to s1 is: are you?
after erase to s2 is:

s1.erase(s1.begin(), s1.begin()+3)表明删除 s1 串的前三个字符,所以结果是 s1 变为"are you?"; s2.erase(s2.begin(), s2.end())表明删除整个 s2 字符串,故没有显示结果。

5.5 字符串比较

主要是依据 ASCII 值来比较大小。若字符串 s1"大于"s2,表明两者相比较时遇到了第一对不同的字符,字符串 s1 中第一个不同的字符比字符串 s2 中同样位置的字符在 ASCII 表中的位置更靠后。

C++ STL 提供了多种字符串比较方法,它们各具特色。其中最简单的就是使用非成员的重载运算符函数 operator==、operator!=、operator>、operator<、operator>=和

第 5 章 字符串

operator<=。

【例 5.7】 字符串比较函数示例。

```
//文件名: e5_7.cpp
#include <string>
#include <iostream>
using namespace std;
int main(int argc, char * argv[])
{
    string s1="this";
    string s2="that";
    if(s1>s2) cout<<"s1>s2"<<endl;
    if(s1==s2) cout<<"s1=s2"<<endl;
    if(s1<s2) cout<<"s1<s2"<<endl;
    return 0;
}
```

可以看出，直接运用重载运算符就可以了，非常方便。s1、s2 前两个字符是相同的，第三个字符比较时由于 i 的 ASCII 大于 a 的 ASCII，因此 s1>s2。

5.6 综合示例

下面列举一些常用功能的简单示例，对一些前文未讲到的知识点做了必要的说明。

【例 5.8】 整型数据与字符串互相转化。

```
//文件名: e5_8.cpp
#include<string>
#include<iostream>
#include<sstream>
using namespace std;
int main(int argc, char * argv[])
{
    //将整型 10 转化成字符串"10"
    int n1=10;
    string s1;
    stringstream os1;
    os1<<n1;
    os1>>s1;
    cout<<"整型 10 转化成字符串 10: "<<s1<<endl;

    //将字符串"123"转化成整型 123
```

```
        int n2=0;
        string s2="123";
        stringstream os2;
        os2<<s2;
        os2 >>n2;
        cout<<"字符串 123 转化成整型 123: "<<n2<<endl;
        return 0;
}
```

执行结果为:

```
整型 10 转化成字符串 10: 10
字符串 123 转化成整型 123: 123
```

可知在 STL 中,实现基本数据类型与字符串的相互转化,用流类如 stringstream 做中间媒介是一个很好的思路。stringstream 是输入输出流,利用输出流功能先填充输出缓冲区,再利用输入流功能依次读缓冲区,赋值给相应的变量。

【例 5.9】 利用 STL 拆分字符串。

例如,有字符串"How are you?"按空格拆分,应得到三个子串:"How","are","you"。

方法 1:

```
//文件名:e5_9_1.cpp
#include<iostream>
#include<string>
using namespace std;
int main(int argc, char * argv[])
{
        string strText="How are you? ";              //源串
        string strSeparator=" ";                     //按空格拆分
        string strResult;                            //拆分结果串
        int size_pos=0;                              //拆分子串结束位置
        int size_prev_pos=0;                         //拆分子串起始位置

        while((size_pos=strText.find_first_of(strSeparator ,size_pos))!=string::npos)
                                                     //找到子串
        {
            strResult=strText.substr(size_prev_pos , size_pos-size_prev_pos);
                                                     //取子串
            cout<<"string="<<strResult<<endl;
            size_prev_pos=++size_pos;     //下一子串起始位置、结束位置=当前子串结束位置+1
        }

        if(size_prev_pos!=strText.size())            //判断有无最后一个子串
```

```
        strResult=strText.substr(size_prev_pos , size_pos-size_prev_pos);
        cout<<"string="<<strResult<<endl;
    }
    return 0;
}
```

执行结果为：

string=how
string=are
string=you?

主要思路分析如下。

(1) 在搜索每一个符合条件的子串前，先使子串起始位置等于结束位置，然后按照 find_firsr_of 函数查找相应子串，若有，则子串结束位置不等于起始位置，再利用 substr 函数取之间的字符，即得子串值。

(2) 程序中的 if 语句是判断最后一个子串的边界条件的，若起始位置等于该字符串的长度，表明已无字符串可取，否则还有一个子串。例如本例中的"How are you?"，当前两个串("How","are")取完后，size_prev_pos＝size_pos＝8，但最后一个串"you?"已搜索不到空格字符，因此跳出 while 循环，if 语句条件成立，可取到子串"you?"。当然，若源串是"How are you?"，结尾多一空格，则所有的子串均可在 while 条件语句中搜索到，if 语句条件不成立，条件体也就不执行了。

方法2：

```
//文件名：e5_9_2.cpp
#include<iostream>
#include<sstream>
#include<string>
using namespace std;
int main(int argc, char * argv[])
{
    string strResult="";
    string strText="How are you";
    istringstream istr(strText);              //用字符串输入流封装字符串

    while(! istr.eof())                       //当非字符串输入流末尾
    {
        istr>>strResult;                      //读输入流并给变量赋值
        cout<<"string="<<strResult<<endl;
    }
    return 0;
}
```

巧妙地把静态字符串进一步封装成 istringstream 对象,变为动态,再通过提取符动态拆分字符串。这种"静态-动态"思维值得借鉴。但是可能有同学会问:这种方法只能按空格拆分,若按其他字符比如","拆分该怎么办呢? 其实,只要用 getline 函数代替提取符 >> 就可以了,因为该函数最后一个参数可以设置拆分字符,代码如下:

```
#include<iostream>
#include<sstream>
#include<string>
using namespace std;
int main(int argc, char * argv[])
{
    string strResult="";
    string strText="How,are,you";
    istringstream istr(strText);              //用字符串输入流封装字符串

    while(! istr.eof())                       //当非字符串输入流末尾
    {
        getline(istr, strResult, ',');        //读输入流并给变量赋值
        cout<<"string="<<strResult<<endl;
    }
    return 0;
}
```

【例 5.10】 C++ STL 实现的 string trim()功能。

即去掉字符串两端的空格,STL 中没有单独的 trim 函数可用,主要是应用 erase、find_first_not_of、find_last_not_of 函数。

```
//文件名:e5_10.cpp
#include<string>
#include<iostream>
using namespace std;

int main(int argc, char * argv[])
{
    string s=" hello ";
    s.erase(0, s.find_first_not_of(" "));     //删除左空格
    s.erase(s.find_last_not_of(" ")+1);       //删除右空格
    cout<<s;
    return 0;
}
```

【例 5.11】 对 string 类的进一步封装。

string 类中提供了丰富的函数,可视为抽象出的字符串功能的最小集。在实际的工程

中,对于所需要的字符串处理的各个功能来说,往往是其中各个函数的有机组合。好的办法是进一步封装 string 类,编制所需要的字符串函数,这样使调用方就可调用所编各函数,完成相应功能。

例如要实现以下功能:
(1) 字符串去空格;
(2) 字符串转化成整型数;
(3) 整型数转化成字符串。

编制的功能类如下:

```cpp
//文件名:e5_11.cpp
#include<string>
#include<sstream>
#include<iostream>
using namespace std;
class ext_string: public string
{
public:
    int GetInt(string strText)                          //把字符串转化为整型
    {
        int n=0;
        stringstream os;
        os<<strText;
        os>>n;
        return n;
    }
    string GetString(int n)                             //把整型转化为字符串
    {
        string s;
        stringstream os;
        os<<n;
        os>>s;
        return s;
    }
    string GetTrim(string strText)                      //去字符串两侧空格
    {
        strText.erase(0, strText.find_first_not_of(" ")); //删除左空格
        strText.erase(strText.find_last_not_of(" ")+1);   //删除右空格
        return strText;
    }
};
int main(int argc, char * argv[])
{
```

```
    ext_string extstring;
    int n=extstring.GetInt("123");              //将串"123"转化为整型数
    string s1=extstring.GetString(456);         //将整型数转化为字符串
    string s2=extstring.GetTrim(" hello ");     //去两侧空格
    cout<<"The string '123' convert to integer is: "<<n<<endl;
    cout<<"The integer 456 convert to string is:"<<s1<<endl;
    cout<<"The string 'hello' erase space is:"<<s2<<endl;
    return 0;
}
```

程序执行结果为:

```
The string '123' convert to integer is: 123
The integer 456 convert to string is: 456
The string 'hello' erase space is:hello
```

ext_string类中三个函数功能在例5.8~例5.10中均有论述。通过比较得出:由于对字符串做了统一的封装,一方面使得调用者程序非常简洁,层次清晰,若按例5.8~例5.10中方法,把所有功能都写在调用程序中,则显得十分臃肿;另一方面,ext_string类可以作为一个工具类,被工程中的其他人员共享使用。

当然,本示例只是说明了扩展字符串类的设计思想,功能并不完善。

第6章 函数对象

前面已经明确了函数的概念,但是把函数作为对象是程序设计的新思维。STL 是通过重载类中的 operator 函数实现函数对象功能的,不但可以对容器中的数据进行各种各样的操作,而且能够维护自己的状态。因此,与标准 C 库函数相比,函数对象更为通用。

6.1 简介

6.1.1 为何引入函数对象

首先看一段示例,功能是采用 STL 固有 for_each 算法求保存在向量中的整数之和。

```
int sum=0;
void f(int n)
{
    sum+=n;
}

int main()
{
    vector<int>v;
    for(int i=1; i<=100; i++)
    {
        v.push_back(i);
    }
    for_each(v.begin(), v.end(), f);
    printf("sum=%d\n", sum);
    return 0;
}
```

很明显,f(int)是全局普通函数,sum 是求和全局变量,结果是显示 1~100 的总和。实际情况是随着面向对象思想的普及和发展,绝大多数的功能都封装在类中,例如上述的求和过程封装在如下类中。

```
class CSum
{
```

```
private:
    int sum;
public:
    CSum(){sum=0;}
    void f(int n)
    {
        sum+=n;
    }
    int GetSum() {return sum;}
}
```

那么,如何在 STL 中调用 CSum 中的 f(int)函数,推而广之,如何调用所需类中的所需函数就是一个十分关键的问题。进一步加以思考,CSum 是对整型数求和,若是下面的模板参数就更好了。

```
template<class T>
class CSum
{
private:
    T sum;
public:
    CSum(){sum=0;}
    void f(T n)
    {
        sum+=n;
    }
    T GetSum() {return sum;}
}
```

上面仅是关于为什么有函数对象思想的一些假设,主要是提供一种思考方法,与最终实现结果还是有偏差的。但关键思想是一致的,即在函数中调用所需类对象的函数,使程序结构显得非常简洁。

6.1.2 函数对象分类

函数对象是重载了 operator()的类的一个实例,operator()是函数调用运算符。标准 C++库根据 operator()参数个数为 0 个,1 个,2 个加以划分。主要有以下 5 种类型。
- 发生器:一种没有参数且返回一个任意类型值的函数对象,例如随机数发生器。
- 一元函数:一种只有一个任意类型的参数,且返回一个可能不同类型值的函数对象。
- 二元函数:一种有两个任意类型的参数,且返回一个任意类型值的函数对象。
- 一元判定函数:返回 bool 型值的一元函数。

- 二元判定函数:返回 bool 型值的二元函数。

可以看出,STL 中函数对象最多仅适用于两个参数,但这也足以完成相当强大的功能。

6.1.3 简单示例

同样是求整型向量各元素之和,利用函数对象后代码如下所示。

【**例 6.1**】 求整型向量各元素之和。

```
//文件名: e6_1.cpp
class CSum
{
private:
    int sum;
public:
    CSum(){sum=0;}
    void operator()(int n)
    {
        sum+=n;
    }
    int GetSum(){return sum;}
};

int main()
{
    vector<int>v;
    for(int i=1; i<=100; i++)
    {
        v.push_back(i);
    }
    CSum sObj=for_each(v.begin(), v.end(), CSum());
    printf("sum=%d\n", sObj.GetSum());
    return 0;
}
```

对这段代码,主要理解以下几点。

(1) 必须重载 operator 函数。这是实现函数对象功能最重要的环节,不能随便写,因此 6.1.1 节中 CSum 类中的 f(int)函数修改为 operator()(int),而函数体的内容不变。

(2) 函数对象调用方式。直接采用构造函数形式调用,如本例中 for_each(v.begin(), v.end(), CSum())中的第三个参数 CSum()。也就是说,对本例而言,STL 知道 CSum()对应着 CSum 类中重载的 operator 函数,具体有几个参数呢?由于 for_each 函数一次只能迭代出一个整型数,因此 STL 知道每迭代一次整型数都要执行一次 CSum 中的 operator()(int)函数。

(3) CSum sObj 用来接收 for_each 迭代函数对象的最终结果值。

6.2 一元函数

STL 中一元函数基类是一个模板类,其原型如下:

```
template<class _A, class _R>
struct unary_function
{
    typedef _A argument_type;
    typedef _R result_type;
};
```

它有两个模板参数,_A 是输入参数,_R 是返回类型,且这两个参数的类型是任意的,因此它的动态特性非常强。例如,在 6.1.3 节简单示例中,编制了求整型向量元素和的功能类 CSum,如果求浮点向量的元素和,则必须编制新的函数对象功能类。如果应用一元函数,则问题变得非常简单,代码如下。

【例 6.2】 利用一元函数求向量各元素之和。

```
//文件名:e6_2.cpp
#include<iostream>
#include<algorithm>
#include<functional>
#include<vector>
using namespace std;

template<class _inPara, class _outPara>
class CSum:public unary_function<_inPara, _outPara>
{
public:
    _outPara sum;
    CSum(){sum=0;}
    void operator()(_inPara n)
    {
        sum+=n;
    }
    _outPara GetSum(){return sum;}
};
int main()
{
    vector<int>v;
    for(int i=1; i<=100; i++)
```

```
    {
        v.push_back(i);
    }
    CSum<int,int>sObj=for_each(v.begin(), v.end(), CSum<int, int>());
    cout<<"sum(int)="<<sObj.GetSum()<<endl;

    vector<float>v2;
    float f=1.3f;
    for(int i=1; i<=99; i++)
    {
        v2.push_back(f);
        f+=1.0f;
    }
    CSum<float,float>sObj2=for_each(v2.begin(), v2.end(), CSum<float, float>());
    cout<<"sum(float)="<<sObj2.GetSum()<<endl;
    return 0;
}
```

主要理解以下几点。

(1) 应用 STL 模板一元函数必须从 unary_function 基类派生。例如本例中的 CSum 类。

(2) 加深对一元函数模板类模板参数的理解。例如本例中的_inPara 表示 operator 函数的参数类型,所以写作 void operator()(_inPara n);_outPara 表示返回值的类型,因此返回值变量 sum 应定义成_outPara 类型。由于这两个参数在本例中都是动态传进去的,因此在类定义的前面要加上 template<class _inPara, class _outPara>。

(3) 对调用模板函数对象方式的理解。例如本例中 CSum<int,int> sObj=for_each(v.begin(), v.end(), CSum<int, int>())表明是对整型向量元素求和,CSum<int,int>表明 CSum 是一个模板类,两个动态参数都是整型数;CSum<float,float>sObj2=for_each(v2.begin(), v2.end(), CSum<float, float>())表明是对浮点向量元素求和,两个动态参数都是浮点型。可以看出对整型向量求和、浮点向量求和都是由 CSum 函数对象类完成的,当然也可用 CSum 求其他数据类型的和,从中可以体会出 STL 中一元函数模板类功能的强大。

那么,本例中累加和均是从 0 开始计算的,如果现在要求从某个基数开始进行累加向量中的各元素值,应该如何修改 CSum 类的代码呢?其实通过把累加基数通过构造函数传进去即可。CSum 类修改如下:

```
template<class _inPara, class _outPara>
class CSum:public unary_function<class _inPara, class _outPara>
{
public:
```

```
    _outPara sum;
    CSum(_outPara init){sum=init;}
    void operator()(_inPara n)
    {
        sum+=n;
    }
    _outPara GetSum(){return sum;}
};
```

若调用代码为 for_each(v. begin()，v. end，CSum<int,int>(100))，则表明以 100 为累加基数，累加整型向量的各元素值。

6.3 二元函数

STL 中二元函数基类是一个模板类，原型如下。

```
template<class Arg1, class Arg2, class Result>
struct binary_function {
    typedef Arg1 first_argument_type;
    typedef Arg2 second_argument_type;
    typedef Result result_type;
};
```

它有三个模板参数，Arg1、Arg2 是输入参数，Result 是返回类型，且这三个参数的类型是任意的，因此它的动态特性非常强。例如若按学生成绩升序排列，利用二元函数后程序代码如下所示。

【例 6.3】 利用二元函数使学生成绩升序排列。

```
//文件名：e6_3.cpp
#include<functional>
#include<algorithm>
#include<iostream>
#include<vector>
#include<string>
#include<iterator>
using namespace std;
class Student
{
public:
    string name;
    int grade;
public:
```

```cpp
    Student(string name, int grade)
    {
        this->name=name;
        this->grade=grade;
    }
    bool operator< (const Student& s) const
    {
        return grade<s.grade;
    }
};

ostream& operator<< (ostream& os, const Student& s)
{
    os<<s.name<<"\t"<<s.grade<<"\n";
    return os;
}
template<class _inPara1, class _inPara2>
class binary_sort:public binary_function<_inPara1, _inPara2, bool>
{
public:
    bool operator()(_inPara1 in1, _inPara2 in2)
    {
        return in1<in2;
    }
};
int main()
{
    Student s1("zhangsan", 60);
    Student s2("lisi", 80);
    Student s3("wangwu", 70);
    Student s4("zhaoliu", 90);
    vector<Student>v;
    v.push_back(s1);
    v.push_back(s2);
    v.push_back(s3);
    v.push_back(s4);

    sort(v.begin(), v.end(), binary_sort<const Student&,const Student&>());
                                                            //按升序排列
```

```
    copy(v.begin(),v.end(),ostream_iterator<Student>(cout,""));    //升序结果输出
    return 0;
}
```

该函数的程序执行流程：主程序中首先把 4 个学生对象依次放入向量 vector 中，当执行排序函数 sort 时，调用二元函数类 binary_sort 中重载的 operator 函数，当执行函数体 in1<in2 时，由于 in1，in2 的参数类型都是 Student&，因此调用 Student 类中重载的 operator< 函数完成两个 Student 引用对象真正的比较功能，随后把返回值依次返回，sort 函数根据此返回值决定这两个 Student 对象是否交换。

6.4 系统函数对象

STL 提供了大量内建的函数对象，主要包含算术类、关系运算类和逻辑运算类三种函数对象，具体如表 6.1 所示。

表 6.1 系统函数对象分类

	名 称	类 型	结果描述
算术类函数对象	加法：plus<T>	二元函数	arg1+arg2
	减法：minus<T>	二元函数	arg1-arg2
	乘法：multiples<T>	二元函数	arg1 * arg2
	除法：divides<T>	二元函数	arg1/arg2
	模取：modules<T>	二元函数	arg1%arg2
	否定：negate<T>	一元函数	-arg1
关系运算类函数对象	等于：equal_to<T>	二元函数	arg1==arg2
	不等于：not_equal_to<T>	二元函数	arg1!=arg2
	大于：greater<T>	二元函数	arg1>arg2
	大于等于：greater_equal<T>	二元函数	arg1>=arg2
	小于：less<T>	二元函数	arg1<arg2
	小于等于：less_equal<T>	二元函数	arg1<=arg2
逻辑运算类函数对象	与：logical_and<T>	二元函数	arg1&&arg2
	或：logical_or<T>	二元函数	arg1||arg2
	非：logical_not<T>	二元函数	!arg1

从表 6.1 可以看出，除了算术类函数对象中 negate 是一元函数外，其余均是二元函数，那为什么只有一个参数<T>呢？其实很简单，这是由于此二元函数中两个输入参数的类

型相同,仅定义一个类型就可以了,而且也能够推测出返回值类型来。对算术类函数对象而言,输入参数、返回值都是类型 T;对于关系类函数对象而言,输入参数类型是 T,返回值类型是布尔;对于逻辑运算类函数对象而言,输入参数类型是 T,返回值类型是布尔。

6.4.1 算术类函数对象

主要包括加法、减法、乘法、除法、模取及否定操作,下面示例演示了一般用法。

【例 6.4】 算术类基本函数对象使用示例。

```cpp
//文件名:e6_4.cpp
#include<functional>
#include<iostream>
using namespace std;
int main()
{
    //以下产生一些算术类函数对象实体
    plus<int>plusObj;
    minus<int>minusObj;
    multiplies<int>mulObj;
    divides<int>divObj;
    modulus<int>modObj;
    negate<int>negObj;
    //以下运用上述对象,履行函数功能
    cout<<plusObj(2,4)<<endl;
    cout<<minusObj(2,4)<<endl;
    cout<<mulObj(2,4)<<endl;
    cout<<divObj(2,4)<<endl;
    cout<<modObj(5,3)<<endl;
    cout<<negObj(4)<<endl;
    //以下直接以函数对象的临时对象履行函数功能
    //语法分析:function<T>()是一个临时对象,调用 operator 运算符
    cout<<plus<int>()(2,4)<<endl;
    cout<<minus<int>()(2,4)<<endl;
    cout<<multiplies<int>()(2,4)<<endl;
    cout<<divides<int>()(2,4)<<endl;
    cout<<modulus<int>()(5,3)<<endl;
    cout<<negate<int>()(4)<<endl;
    return 0;
}
```

由该例可知,对常规通用数据类型 char、int、float、string 而言,可以直接按上述写法进

行。但是对非常规数据类型,则必须重载类中的各个 operator 算术运算符。例如,若有复数类 Complex,用 plus<Complex>完成两个复数加法的代码如下所示。

【例 6.5】 非常规数据类型基本函数对象使用示例。

```cpp
//文件名：e6_5.cpp
#include<functional>
#include<iostream>
using namespace std;
class Complex
{
public:
    float real;                                    //复数实部
    float virt;                                    //复数虚部
public:
    Complex()
    {
        this->real=0.0f;
        this->virt=0.0f;
    }
    Complex(float real, float virt)
    {
        this->real=real;
        this->virt=virt;
    }

    Complex operator+ (const Complex& c) const
    {
        Complex v;
        v.real=real+c.real;
        v.virt=virt+c.virt;
        return v;
    }
};
int main()
{
    Complex c1(1.0f, 2.0f);
    Complex c2(3.0f, 4.0f);
    Complex c3=c1+c2;
    Complex c4=plus<Complex>()(c1,c2);
    cout<<c3.real<<"+"<<c3.virt<<"i"<<endl;
    cout<<c4.real<<"+"<<c4.virt<<"i"<<endl;
    return 0;
}
```

程序的功能是计算两个复数 1+2i 与 3+4i 的和,主要理解点如下。

(1) 主程序中 Complex c3＝c1＋c2 流程的理解。当运行到该行时,直接调用 Complex

类中 operator＋重载函数，且当前对象是 c1(1＋2i)，传入对象是 c2(3＋4i)，在函数体中得出相加后的实部值 4、虚部值 6，并返回该复数。

（2）主程序中 Complex c4＝plus＜Complex＞()(c1，c2)流程的理解。当运行到该行时，先执行二元函数 plus 类中重载的 operator 函数。plus 类原型如下所示。

```
template<class T>
struct plus:public binary_function<T,T,T>
{
    T operator()(const T& x, const T&y) const
    {
        return x+y;
    }
}
```

由于 x，y 是非常规数据类型，当执行到函数体 x＋y 时，则执行 Complex 类中重载的 operator＋运算符函数，完成两复数相加，并把结果依次返回。结果也是 4＋6i。

很明显，复数 c3、c4 得到的结果是一样的，况且 c3＝c1＋c2 比 c4＝plus＜Complex＞()(c1，c2)要简洁得多，那么定义二元函数 plus 不就没有用处了吗？其实，单独用二元函数实现相应功能优势是不大的，必须与 STL 算法结合才能体现出它的好处来。仍然以上述的复数类为例，仅把主函数改为：

```
int main()
{
    Complex c1(1.0f, 2.0f);
    Complex c2(3.0f, 4.0f);
    Complex c3=c1+c2;
    Complex c4=plus<Complex>()(c1, c2);
    Complex c;
    vector<Complex>v;
    v.push_back(c1);
    v.push_back(c2);
    v.push_back(c3);
    v.push_back(c4);
    Complex result=accumulate(v.begin(), v.end(), c, plus<Complex>());
    cout<<result.real<<"+"<<result.virt<<"i";
    return 0;
}
```

功能是先定义 4 个复数，再保存至向量中，通过向量求这 4 个复数的和。STL 数值函数 accumulate 第 4 个参数表明调用的是一个系统二元函数 plus，是加法运算。在这里体现出了系统二元函数的优越性。

6.4.2 关系运算类函数对象

主要包括等于、不等于、大于、大于等于、小于、小于等于 6 种运算，下面示例演示了一般用法。

【例 6.6】 关系运算类基本函数对象使用示例。

```
//文件名：e6_6.cpp
#include<functional>
#include<iostream>

using namespace std;
int main()
{
    //以下产生一些关系运算类函数对象实体
    equal_to<int>equalObj;
    not_equal_to<int>notequalObj;
    greater<int>greatObj;
    greater_equal<int>greatequalObj;
    less<int>lessObj;
    less_equal<int>lessequalObj;
    //运用上述对象执行功能
    cout<<equalObj(2, 4)<<endl;
    cout<<notequalObj(2, 4)<<endl;
    cout<<greatObj(2, 4)<<endl;
    cout<<greatequalObj(2, 4)<<endl;
    cout<<lessObj(2, 4)<<endl;
    cout<<lessequalObj(2, 4)<<endl;
    //以下直接以临时对象执行函数功能
    cout<<equal_to<int>()(2, 4)<<endl;
    cout<<not_equal_to<int>()(2, 4)<<endl;
    cout<<greater<int>()(2, 4)<<endl;
    cout<<greater_equal<int>()(2, 4)<<endl;
    cout<<less<int>()(2, 4)<<endl;
    cout<<less_equal<int>()(2, 4)<<endl;
    return 0;
}
```

由该例可知，对常规通用数据类型 char、int、float、string 而言，可以直接按上述写法进行。但是对非常规数据类型，则必须重载类中的各个 operator 关系运算符。例如，若有复数类 Complex，用 equal_to<Complex>完成两个复数相等比较的代码如下所示。

【例 6.7】 利用二元函数比较两复数是否相等问题。

```cpp
//文件名：e6_7.cpp
#include<functional>
#include<iostream>
using namespace std;
const class Complex
{
public:
    float real;                                    //复数实部
    float virt;                                    //复数虚部
public:
    Complex()
    {
        this->real=0.0f;
        this->virt=0.0f;
    }
    Complex(float real, float virt)
    {
        this->real=real;
        this->virt=virt;
    }
    bool operator==(const Complex& c) const
    {
        return ((real==c.real)&&(virt==c.virt));
    }

};
int main()
{
    Complex c1(1.0f, 2.0f);
    Complex c2(3.0f, 4.0f);
    Complex c3(1.0f, 2.0f);
    cout<<equal_to<Complex>()(c1, c2)<<endl;
    cout<<equal_to<Complex>()(c1, c3)<<endl;
    return 0;
}
```

主要理解 equal_to 的流程。当执行到 equal_to<Complex>()(c1, c2)时，首先调用二元函数类 equal_to 中重载的 operator==运算符函数。由于 Complex 是非常规数据类型，之后接着调用 Complex 中重载的 operator==运算符函数，完成真正的复数比较，布尔值依次返回。

6.4.3 逻辑运算类函数对象

主要有与、或、非三种运算。基本用法见例 6.8。

【例 6.8】 逻辑运算类函数对象基本用法。

```cpp
//文件名：e6_8.cpp
#include<functional>
#include<iostream>
using namespace std;
int main()
{
    //产生一些函数对象实体
    logical_and<int>andObj;
    logical_or<int>orObj;
    logical_not<int>notObj;
    //运用上述对象执行函数功能
    cout<<andObj(true, true)<<endl;                    //1
    cout<<orObj(true, false)<<endl;                    //1
    cout<<notObj(true)<<endl;                          //0
    //利用临时对象执行函数功能
    cout<<logical_and<int>()(3<5, 6<9)<<endl;
    cout<<logical_or<int>()(3<5, 6>9)<<endl;
    cout<<logical_not<int>()(3<5)<<endl;
    return 0;
}
```

6.4.4 函数适配器

函数适配器主要有以下几方面的优点。

（1）可以形成复杂的、有语义的表达式。上文讲述的算术类、关系类、逻辑运算类函数对象功能已经很强大了，但某些时候仍显不足，例如，求某整型数组 a 中大于 10 的数的个数。很明显可知：若用系统提供的函数对象，那么一定是 greater，可是下述写法又无法体现出待比较的数 10：

  int n=count_if(a, a+sizeof(a)/sizeof(int), greater<int>())

这种写法是错误的，利用适配器可以很好地解决这种问题。

（2）可以调用类中普通的成员函数。我们熟知：STL 中绝大多数算法归根结底是调用功能类中重载的 operator 运算符来实现的，然而，功能类中还有许多普通的成员函数，STL 本身不能直接调用，必须经过适配器转换才可调用。

函数适配器分类如表 6.2 所示。

表 6.2 函数适配器分类表

类型	函数	原型	说明
绑定	bind1st()	template<class Pred, class T> binder1st<Pred> **bind1st**(const Pred& pr, const T& x);	(1) Pred 是二元函数。 (2) 二元函数对象第一个参数绑定为 x。 (3) 返回值相当于 binder1st<Pred>(pr, Pred::first_argument_type(x))
绑定	bind2nd()	template<class Pred, class T> binder2nd<Pred> **bind2nd**(const Pred& pr, const T& y);	(1) Pred 是二元函数。 (2) 二元函数对象第二个参数绑定为 y。 (3) 返回值相当于 binder2nd<Pred>(pr, Pred::second_argument_type(y))
取反	not1	template<class Pred> unary_negate < Pred > **not1** (const Pred& pr);	(1) Pred 是一元函数。 (2) 返回值相当于 unary_negate<Pred>(pr)
取反	not2	template<class Pred> binary_negate < Pred > **not2** (const Pred& pr);	(1) Pred 是二元函数。 (2) 返回值相当于 binary_negate<Pred>(pr)
成员函数适配器	mem_fun	template<class R, class T> mem_fun_t< R, T > **mem_fun** (R (T::* pm) ());	调用类 T 中的成员函数
成员函数适配器	mem_fun_ref	template<class R, class T> mem_fun_ref_t<R, T> **mem_fun_ref**(R (T:: * pm) ());	调用类 T 中的成员函数

普通函数适配器	ptr_fun	template<class Arg, class Result>
		  pointer_to_unary_function<Arg, Result>
		  ptr_fun(Result (*pf)(Arg));
			把全局函数进一步封装成一元函数

普通函数适配器	ptr_fun	template< class Arg1, class Arg2, class Result>
		  pointer_to_binary_function<Arg1, Arg2, Result>
		  ptr_fun(Result(*pf)(Arg1, Arg2));
			把全局函数进一步封装成二元函数

【例 6.9】 绑定、取反适配器基本用法。

```
//文件名:e6_9.cpp
#include<functional>
#include<algorithm>
#include<iostream>
#include<iterator>
using namespace std;
int main()
{
    int a[]={1,3,5,7,9,8,6,4,2,0};
    int nCount=count_if(a, a+sizeof(a)/sizeof(int), bind2nd(less<int>(), 4));
```

```
        cout<<nCount<<endl;                    //4
        nCount=count_if(a, a+sizeof(a)/sizeof(int), bind1st(less<int>(), 4));
        cout<<nCount<<endl;                    //5
        nCount=count_if(a, a+sizeof(a)/sizeof(int), not1(bind2nd(less<int>(), 4)));
        cout<<nCount<<endl;                    //6
        nCount=count_if(a, a+sizeof(a)/sizeof(int), not1(bind1st(less<int>(), 4)));
        cout<<nCount<<endl;                    //5
        sort(a, a+sizeof(a)/sizeof(int), not2(less<int>()));
        copy(a, a+sizeof(a)/sizeof(int), ostream_iterator<int>(cout, " "));
                                               //9 8 7 6 5 4 3 2 1 0
        return 0;
    }
```

着重理解以下的知识点。

(1) bind2nd(less<int>(),4)：less<int>()本身是一个二元函数对象,相当于普通函数 bool less(T x,T y){ return x<y;}；而 bind2nd(less<int>(),4))相当于普通函数 bool less(T x, int n=4){return x<4},bind2nd 的作用是使 less 二元函数的第二个参数绑定为整型 4,因此相当于把二元函数降低为一元函数。count_if 语句中是求数组 a 中小于 4 的数据个数是多少,可知符合条件的数有{1,3,2,0},共 4 个。

(2) bind1st(less<int>(),4)：less<int>()本身是一个二元函数对象,相当于普通函数 bool less(T x,T y){ return x<y;}；而 bind1st(less<int>(),4))相当于普通函数 bool less(int n=4, T x){return 4<x},bind1st 的作用是使 less 二元函数的第一个参数绑定为整型 4,因此相当于把二元函数降低为一元函数。count_if 语句中是求数组 a 中大于 4 的数据个数是多少,可知符合条件的数有{5,7,9,8,6},共 5 个。

(3) not1(bind2nd(less<int>(), 4))：bind2nd(less<int>(), 4)相当于普通函数 bool less(T x, int n=4){return x<4;},not1 后相当于 bool less(T x, int n=4){return !(x<4);}。语义上相当于"不小于 4",即求不小于 4 的数据个数是多少,可知符合条件的数有{5,7,9,8,6,4},共有 6 个。

(4) not1(bind1st(less<int>(), 4))：bind1st(less<int>(), 4)相当于普通函数 bool less(int n=4, T x){return 4<x;},not1 后相当于 bool less(int n=4, T x){return !(4<x);}。语义上相当于"不大于 4",即求不大于 4 的数据个数是多少,可知符合条件的数有{1,3,4,2,0},共 5 个。

(5) not2(less<int>())：less<int>()相当于普通函数 bool less(T x, T y){return x<y;},not2 后相当于 bool less(T x, T y){return !(x<y);},即求数组 a 的降幂排序序列。

有一点需要注意,not1 是针对一元函数而言的,可以直接操作一元函数,也可通过 bind2nd、bind1st 把二元函数降为一元函数后再使用;not2 是针对二元函数而言的,直接作用即可。

【例 6.10】 成员函数适配器基本用法。

//文件名：e6_10.cpp

```cpp
#include<functional>
#include<algorithm>
#include<iostream>
#include<vector>
#include<string>
using namespace std;
class Student
{
    string strNO;                              //学号
    string strName;                            //姓名
public:
    Student(string strNO,string strName):strNO(strNO),strName(strName){}
    bool show()                                //学生信息显示函数
    {
        cout<<strNO<<":"<<strName<<endl;
        return true;
    }
};
void main()
{
    //针对 mem_fun_ref 程序段
    Student s1("1001","zhangsan");
    Student s2("1002","lisi");
    vector<Student>v;
    v.push_back(s1);
    v.push_back(s2);
    for_each(v.begin(), v.end(), mem_fun_ref(Student::show));

    //针对 mem_fun 程序段
    Student * ps1=new Student("1003","wangwu");
    Student * ps2=new Student("1004","zhaoliu");
    vector<Student * >pv;
    pv.push_back(ps1);
    pv.push_back(ps2);
    for_each(pv.begin(), pv.end(), mem_fun(Student::show));
}
```

该段程序的主要功能是添加学生信息对象向量,并调用 Student 类中的 Show 函数,屏幕上显示学生信息。主要理解点如下。

(1) mem_fun_ref、mem_fun 的区别:若集合是基于对象的,形如 vector<Student>,则用 mem_fun_ref;若集合是基于对象指针的,形如 vector<Student * >,则用 mem_fun。

(2) 以下调用写法是错误的:for_each(v.begin(), v.end(), Student::show),当在 STL 算法中调用成员函数时,一般要通过 mem_fun_ref 或 mem_fun 转换后才可以应用。

【例 6.11】 普通函数适配器基本用法。

```cpp
//文件名：e6_11.cpp
#include<functional>
#include<algorithm>
#include<iostream>
using namespace std;
bool f(int x)
{
    return x>3;
}
bool g(int x, int y)
{
    return x>y;
}
int main()
{
    int a[]={2,5,3,7,1,9,8,0,6};
    int nSize=sizeof(a)/sizeof(int);
    int nCount=count_if(a, a+nSize, f);                      //第 1 个 count_if
    cout<<nCount<<endl;                                      //5
    nCount=count_if(a, a+nSize, ptr_fun(f));                 //第 2 个 count_if
    cout<<nCount<<endl;                                      //5
    nCount=count_if(a, a+nSize, bind2nd(ptr_fun(g), 3));     //第 3 个 count_if
    cout<<nCount<<endl;                                      //5
    nCount=count_if(a, a+nSize, bind2nd(ptr_fun(g), 5));     //第 4 个 count_if
    cout<<nCount<<endl;                                      //4
    return 0;
}
```

首先分别定义了输入一个变量、两个变量的普通函数，主程序是显示数组中大于3、大于6的数据个数。主要理解点如下。

(1) 第1个、第2个 count_if 语句中调用了普通函数 f(x)。f(x)中把比较条件固定死了，即只能求大于3的数据个数。第2个 count_if 中普通函数用适配器 ptr_fun(f)加以修饰，第1个 count_if 中直接使用 f 做参数，但它们的输出结果是相同的，都是5。所以对一个参数的普通函数而言，使用 ptr_fun 适配器与否并不能明显看出它的优势。

(2) 第3个、第4个 count_if 语句中调用了普通函数 g(x,y)。用 bind2nd(ptr_fun(g), 3)等效于调用 g(x, 3)，用 bind2nd(ptr_fun(g), 5)等效于调用 g(x,5)，表明分别求大于3、大于5的数据个数，它是通过先用 ptr_fun(g)适配器修饰，再通过绑定适配器 bind2nd 把 g(x,y)函数的第2个参数固定为3、5来实现的，这样的好处是动态性比较强。因此，这种情况下把二元普通函数用 ptr_fun 适配是必需的。

6.5 综合示例

【例 6.12】 编程求圆和长方形的面积。

```cpp
//文件名: e6_12.cpp
#include<functional>
#include<algorithm>
#include<iostream>
#include<vector>
using namespace std;
class Shape                              //在基类中定义求面积的多态虚函数
{
public: virtual bool ShowArea()=0;
};
class Circle:public Shape
{
    float r;                             //半径
public:
    Circle(float r):r(r){}
    bool ShowArea(){                     //重载多态虚函数
        cout<<3.14f*r*r<<endl;
        return true;
    }
};
class Rectangle:public Shape
{
    float width, height;
public:
    Rectangle(float width, float height):width(width),height(height){}
    bool ShowArea() {                    //重载多态虚函数
        cout<<width*height<<endl;
        return true;
    }
};
class AreaCollect                        //各种形状对象的指针集合类
{
    vector<Shape *>v;
public:
    bool Add(Shape * pShape)             //添加形状对象指针函数
    {
        v.push_back(pShape);
```

```
            return true;
        }
        bool ShowEachArea()                    //显示各形状对象面积函数
        {
            for_each(v.begin(), v.end(), mem_fun(Shape::ShowArea));
            return true;
        }
};
void main()
{
    AreaCollect contain;
    Shape * pObj1=new Circle(10.0f);
    Shape * pObj2=new Rectangle(10.0f, 20.0f);
    contain.Add(pObj1);
    contain.Add(pObj2);
    contain.ShowEachArea();
}
```

主要理解点如下。

(1) 设计思想的理解。由于求两种形状的面积,因此对基本类而言最好用多态去实现:在基类 Shape 中定义了纯虚函数 bool ShowArea()。Circle(圆类)、Rectangle(矩形类)均是从 Shape 派生的,并重载了多态函数 ShowArea。AreaCollect 是一个维护各种形状对象的集合类,本例中仅包含各种形状对象指针的添加函数 Add,各种形状对象的面积显示函数 ShowEachArea。测试 main 函数中向集合类中添加了一个圆对象指针,一个长方形对象指针,并调用面积显示函数,把面积显示出来。

(2) 由于采用了多态思想,集合类 AreaCollect 中成员变量必须定义成 vector<Shape*>形式的,不能定义成 vector<Shape>的形式。

(3) 集合类 AreaCollect 成员函数 ShowEachArea()函数中应用了 STL 算法 for_each,其中的第三个参数采用了成员函数适配器 mem_fun,这是由于集合 v 是基于各种形状对象指针的。也就是说,在多态应用程序中,mem_fun 适配器会经常用到。

【例 6.13】 假设学生对象集合的索引从 0 开始,依次增 1。要求不改变学生成绩集合中的学生对象顺序,依据成绩升序输出索引。

```
//文件名:e6_13.cpp
#include<functional>
#include<algorithm>
#include<iostream>
#include<vector>
#include<iterator>
#include<string>
using namespace std;
class Student                                                    //学生类
```

```cpp
    public:
        string strNO;                                    //学号
        int nGrade;                                      //成绩
    public:
        Student(string strNO, int nGrade):strNO(strNO), nGrade(nGrade){}
    };
    class StudIndex:public binary_function<int,int,bool>
    {
        vector<Student> &vStud;                          //学生集合对象
    public:
        StudIndex(vector<Student>&vStud):vStud(vStud){}
        bool operator() (int a, int b)
        {
            return vStud.at(a).nGrade<vStud.at(b).nGrade;
        }
    };
    int main()                                           //仿真测试程序
    {
        Student s1("1001", 70);
        Student s2("1002", 60);
        Student s3("1003", 80);
        Student s4("1004", 74);
        vector<Student>v;                                //添加 4 个学生对象到集合 v
        v.push_back(s1);
        v.push_back(s2);
        v.push_back(s3);
        v.push_back(s4);

        vector<int>vIndex;                               //形成 4 个学生的初始索引号集合
        vIndex.push_back(0);
        vIndex.push_back(1);
        vIndex.push_back(2);
        vIndex.push_back(3);
        sort(vIndex.begin(), vIndex.end(), StudIndex(v));
        copy(vIndex.begin(),vIndex.end(),ostream_iterator<int>(cout," "));
                                                         //1 0 3 2
        return 0;
    }
```

主要理解的知识点如下。

(1) 学生基本信息类 Student 非常简单，成员变量仅包含学号及成绩。

(2) StudIndex 是一个标准的二元函数类。成员变量 vector<Student> &vStud 定义

了学生集合对象的引用变量,它是通过构造函数 StudIndex(vector<Student> &vStud)加以初始化的。运算符函数 bool operator()(int a,int b)是真正被调用的二元函数实现体。

(3) main 是一个测试函数。先定义了 4 个学生对象,把它们依次添加到集合类 v 中;之后索引为 0 开始,依次递增为 1,形成这 4 个学生对象的索引集合 vIndex;最后通过 sort 函数,依据成绩升序信息对索引集合 vIndex 排序,结果为"1 0 3 2",是正确的。

(4) 对 sort(vIndex.begin(),vIndex.end(),StudIndex(v))的理解。可看出是对学生索引集合 vIndex 中各元素排序,并不是对学生集合对象 v 排序,符合题目要求。那么怎么对 vIndex 进行排序呢?根据第三个参数 StudIndex(v)可知,一定是调用了 StudIndex 中重载的 bool operator()(int a, int b)函数,a,b 表示学生对象的索引序号,该函数体内容 return vStud.at(a).nGrade<vStud.at(b).nGrade 返回两个学生对象成绩的比较信息。StudIndex(v)的另一个含义是调用 StudIndex 类的构造函数 StudIndex(vector<Student> &vStud),把学生集合对象 v 传入 StudIndex 对象中。

(5) 当然,同学们会发现:若本例中把 class StudIndex:public binary_function<int,int,bool>修改为 class StudIndex,其余都不变,结果也是正确的。这是因为二元函数的两个输入参数,一个输出参数的类型都固定的。若把该类改为如下:

```
template<class T>
class StudIndex:public binary_function<T,T,bool>
{
    vector<Student> &vStud;
public:
    StudIndex(vector<Student> &vStud):vStud(vStud){}
    bool operator() (T a, T b)
    {
        return vStud.at(a).nGrade<vStud.at(b).nGrade;
    }
};
```

两个输入模板参数类型不固定,若想产生正确的结果,必须把测试类中 sort 排序语句所在行改为 sort(vIndex.begin(),vIndex.end(),StudIndex<int>(v)),即把 StudIndex(v)修改为 StudIndex<int>(v)。

第 7 章 通用容器

读者可能都有这样的经历,编制了动态数组类、链表类、集合类和映射类等程序,而且小心地维护着。其实 STL 提供了专家级的几乎我们所需要的各种容器,功能更好,复用性更强,所以开发应用系统应该用 STL 容器类,摈弃自编的容器类,尽管它可能花费了很多的开发时间。

7.1 概述

7.1.1 容器分类

(1) 序列性容器:按照线性排列来存储某类型值的集合,每个元素都有自己特定的位置,顺序容器主要有 vector、deque 和 list。

- vector:就是动态数组。它是在堆中分配内存,元素连续存放,有保留内存,如果减少大小后内存也不会释放。新值大于当前大小时才会再分配内存。对最后元素操作最快(在后面添加删除最快),此时一般不需要移动内存。只有保留内存不够时,才需要对中间和开始处进行添加删除元素操作,这时需要移动内存,如果你的元素是结构或是类,那么移动的同时还会进行构造和析构操作。vector 的一大特点是可直接访问任何元素。
- deque:与 vector 类似,支持随机访问和快速插入删除,它在容器中某一位置上的操作所花费的是线性时间。与 vector 不同的是,deque 还支持从开始端插入、删除数据。由于它主要对前端、后端进行操作,因此也叫做双端队列。
- list:又叫链表,是一种双线性列表,只能顺序访问(从前向后或者从后向前),与前面两种容器类有一个明显的区别就是它不支持随机访问。要访问表中某个下标处的项需要从表头或表尾处(接近该下标的一端)开始循环。

(2) 关联式容器:与前面讲到的顺序容器相比,关联式容器更注重快速和高效地检索数据的能力。这些容器是根据键值(key)来检索数据的,键可以是值也可以是容器中的某一成员。这一类中的成员在初始化后都是按一定顺序排好序的。关联式容器主要有 set、multiset、map 和 multimap。

- set：快速查找，不允许重复值。
- multiset：快速查找，允许重复值。
- map：一对一映射，基于关键字快速查找，不允许重复值。
- multimap：一对多映射，基于关键字快速查找，允许重复值。

(3) 容器适配器：对已有的容器进行某些特性的再封装，不是一个真正的新容器。主要有 stack、queue。

- stack：堆栈类，特点是后进先出。
- queue：队列类，特点是先进先出。

7.1.2 容器共性

容器一般来说都有下列函数。

- 默认构造函数：提供容器默认初始化的构造函数。
- 复制构造函数：将容器初始化为现有同类容器副本的构造函数。
- 析构函数：不再需要容器时进行内存整理的析构函数。
- empty：容器中没有元素时返回 true，否则返回 false。
- max_size：返回容器中最大元素个数。
- size：返回容器中当前元素个数。
- operator＝：将一个容器赋给另一个容器。
- operator＜：如果第一个容器小于第二个容器，返回 true，否则返回 false。
- operator＜＝：如果第一个容器小于或等于第二个容器，返回 true，否则返回 false。
- operator＞：如果第一个容器大于第二个容器，返回 true，否则返回 false。
- operator＞＝：如果第一个容器大于或等于第二个容器，返回 true，否则返回 false。
- operator＝＝：如果第一个容器等于第二个容器，返回 true，否则返回 false。
- operator!＝：如果第一个容器不等于第二个容器，返回 true，否则返回 false。
- swap：交换两个容器的元素。

顺序容器和关联容器共有函数如下。

- begin：该函数有两个版本，返回 iterator 或 const_iterator，返回容器第一个元素迭代器指针。
- end：该函数有两个版本，返回 iterator 或 const_iterator，返回容器最后一个元素后面一位的迭代器指针。
- rbegin：该函数有两个版本，返回 reverse_iterator 或 const_reverse_iterator，返回容器最后一个元素的迭代器指针。
- rend：该函数有两个版本，返回 reverse_iterator 或 const_reverse_iterator，返回容器首个元素前面一位的迭代器指针。
- erase：从容器中清除一个或几个元素。
- clear：清除容器中所有元素。

7.1.3 容器比较

vector(连续的空间存储,可以使用[ ]操作符)快速地访问随机的元素,快速地在末尾插入元素,但是在序列中随机插入、删除元素比较慢。而且如果一开始分配的空间不够的话,有一个重新分配更大空间的过程。

deque(小片的连续,小片间用链表相连,实际上内部有一个 map 的指针,因为知道类型,所以还是可以使用[ ],只是速度没有 vector 快)快速地访问随机的元素,快速地在开始和末尾插入元素,随机地插入、删除元素要慢,空间的重新分配要比 vector 快,重新分配空间后,原有的元素不需要备份。对 deque 的排序操作,可将 deque 先复制到 vector,排序后再复制回 deque。

list(每个元素间用链表相连)访问随机元素不如 vector 快,随机地插入元素比 vector 快,对每个元素分配空间,所以不存在空间不够,重新分配的情况。

set 内部元素唯一,用一棵平衡树结构来存储,因此遍历的时候就排序了,查找也比较快。

map 一对一地映射结合,key 不能重复。

7.2 vector 容器

7.2.1 概述

vector 类称作向量类,它实现了动态的数组,用于元素数量变化的对象数组。像数组一样,vector 类也用从 0 开始的下标表示元素的位置;但和数组不同的是,当 vector 对象创建后,数组的元素个数会随着 vector 对象元素个数的增大和缩小而自动变化。

vector 类常用的函数如下所示。

(1) 构造函数。

- vector():创建一个空 vector。
- vector(int nSize):创建一个 vector,元素个数为 nSize。
- vector(int nSize, const T& t):创建一个 vector,元素个数为 nSize,且值均为 t。
- vector(const vector&):复制构造函数。

(2) 增加函数。

- void push_back(const T& x):向量尾部增加一个元素 x。
- iterator insert(iterator it, const T& x):向量中某一元素前增加一个元素 x。
- void insert(iterator it, int n, const T& x):向量中某一元素前增加 n 个相同元素 x。
- void insert(iterator it, const_iterator first, const_iterator last):向量中某一元素前插入另一个相同类型向量的[first,last)间的数据。

(3) 删除函数。
- iterator erase(iterator it)：删除向量中某一个元素。
- iterator erase(iterator first, iterator last)：删除向量中[first,last)中元素。
- void pop_back()：删除向量中最后一个元素。
- void clear()：删除向量中所有元素。

(4) 遍历函数。
- reference at(int pos)：返回 pos 位置元素的引用。
- reference front()：返回首元素的引用。
- reference back()：返回尾元素的引用。
- iterator begin()：返回向量头指针，指向第一个元素。
- iterator end()：返回向量尾指针，不包括最后一个元素，在其下面。
- reverse_iterator rbegin()：反向迭代器，最后一个元素迭代指针。
- reverse_iterator rend()：反向迭代器，第一个元素之前的迭代指针。

(5) 判断函数。
bool empty() const：向量是否为空，若 true，则向量中无元素。

(6) 大小函数。
- int size() const：返回向量中元素的个数。
- int capacity() const：返回当前向量所能容纳的最大元素值。
- int max_size() const：返回最大可允许的 vector 元素数量值。

(7) 其他函数。
- void swap(vector&)：交换两个同类型向量的数据。
- void assign(int n, const T& x)：设置容器大小为 n 个元素每个元素值为 x。
- void assign(const_iterator first, const_iterator last)：容器中[first,last)中元素设置成当前向量元素。

7.2.2 初始化示例

【例 7.1】 向量初始化方法举例（一般来说有三种方法）。

//文件名：e7_1.cpp
方法 1：

```
#include <vector>
#include <iostream>

class A
{
};
int main(int argc, char * argv[])
```
//无特殊含义，仅是为了配合说明问题

```
    {
        std::vector<int>int_ect;
        std::vector<float>flo_vect;
        std::vector<A>cA_vect;
        std::vector<A*>cpA_vect;

        cout<<"init success!"<<endl;

        return 0;
    }
```

方法 2：

```
#include<vector>
#include<iostream>
using namespace std;

class A                             //无特殊含义,仅是为了配合说明问题
{
};
int main(int argc, char * argv[])
{
    vector<int>int_ect;
    vector<float>flo_vect;
    vector<A>cA_vect;
    vector<A*>cpA_vect;

    cout<<"init success!"<<endl;

    return 0;
}
```

方法 3：

```
#include<vector>
#include<iostream>
using namespace std;

class A                             //无特殊含义,仅是为了配合说明问题
{
};

typedef vector<int>      INT_VECT;
typedef vector<float>    FLO_VECT;
typedef vector<A>        CA_VECT;
```

```
    typedef vector<A *>      CPA_VECT;

    int main(int argc, char * argv[])
    {
        INT_VECT int_ect;
        FLO_VECT flo_vect;
        CA_VECT cA_vect;
        CPA_VECT cpA_vect;

        cout<<"init success!"<<endl;

        return 0;
    }
```

通过比较,可以得出如下结论。

(1) 向量类可以定义基本变量、类、类的指针等。

(2) 若程序中用到标准库,必须包含相应的头文件,本例包含的头文件是<vector>及<iostream>。

(3) 若用了命名空间 using namespace std,则在程序中可直接定义,如方法 2 中 vector<int>int_ect;;否则必须用前缀 std,如方法 1 中 std::vector<int> int_ect;。

(4) 方法 2 与方法 3 的一个主要区别是方法 3 中用 typedef 对 vector 进行了重新定义,它的好处是:一方面主程序简洁,与普通类的用法一致;另一方面程序容易维护。例如,按方法 2 来说,若现在 vector<int>要改成 vector<short>,必须把主程序中所有相关行都修改;若按方法 3,则主程序不动,仅把 typedef vector<int> INT_VECT 改为 typedef vector<short> INT_VECT 即可。

7.2.3 增加及获得元素示例

【例 7.2】 针对基本变量的向量增加及获得元素示例。定义一个整型元素的向量类,先增加两个元素,之后在屏幕上重新显示这两个元素。

```
//文件名: e7_2.cpp
#include<vector>
#include<iostream>
using namespace std;

int main(int argc, char * argv[])
{
    vector<int>int_vec;
    int_vec.push_back(1);                    //通过 push_back 函数添加
    int_vec.push_back(2);
```

```
        int nSize=int_vec.size();                    //获得向量元素数

        cout<<"通过数组方式输出:";                    //通过数组方式输出
        for(int i=0; i<nSize; i++)
        {
            cout<<int_vec[i]<<"\t";
        }
        cout<<endl;                                   //换行

        cout<<"通过获得引用输出:";                    //通过获得引用输出
        for(int i=0; i<nSize; i++)
        {
            int &nValue=int_vec.at(i);
            cout<<nValue<<"\t";
        }
        cout<<endl;

        cout<<"通过迭代器输出:";                      //通过迭代器输出
        vector<int>::iterator int_vec_iter=int_vec.begin();
        while(int_vec_iter!=int_vec.end())
        {
            cout<< * int_vec_iter<<"\t";
            int_vec_iter++;
        }
        cout<<endl;
        return 0;
    }
```

执行程序结果显示如下所示。

通过数组方式输出: 1 2
通过获得引用输出: 1 2
通过迭代器输出: 1 2

可以得出如下主要结论:

(1) 获得向量中的元素值可通过三种常用方式: 直接用向量数组, 获得元素引用, 获得元素的指针。

(2) 迭代器变量 int_vec_iter 表示向量元素的指针, 所以 * int_vec_iter 才表示元素真正的值。

【例 7.3】 针对类的向量增加及获得元素示例。已知类 A 如下, 分别定义元素为 A, A * 的向量类, 增加对应的向量元素, 并在屏幕上重新显示对应的向量元素值 (即显示 A 中的成员变量值 n)。

```cpp
//文件名: e7_3.cpp
#include <vector>
#include <iostream>
using namespace std;
class A
{
public:
    int n;
public:
    A(int n)
    {
        this->n=n;
    }
};

int main(int argc, char * argv[])
{
    /*******************************类A的向量操作*******************************/
    cout<<"第一部分:类A的向量操作"<<endl;
    vector<A>ca_vec;                                            //类A的向量定义
    A a1(1);
    A a2(2);
    ca_vec.push_back(a1);                                       //添加类A的向量元素
    ca_vec.push_back(a2);
    int nSize=ca_vec.size();
    cout<<"通过数组输出:";
    for(int i=0; i<nSize; i++)                                  //通过数组输出
    {
        cout<<ca_vec[i].n<<"\t";
    }
    cout<<"\n通过引用输出:";
    for(int i=0; i<nSize; i++)                                  //通过引用输出
    {
        A& a=ca_vec.at(i);
        cout<<a.n<<"\t";
    }
    cout<<"\n通过迭代器输出:";
    vector<A>::iterator ca_vec_iter=ca_vec.begin();             //通过迭代器输出
    while(ca_vec_iter!=ca_vec.end())
    {
        cout<<(*ca_vec_iter).n<<"\t";
```

```cpp
            ca_vec_iter++;
        }
        cout<<"\n第二部分:类 A 的指针向量操作"<<endl;
        /*****************************类 A 的指针操作*****************************/
        vector<A *>pca_vec;                                    //类 A 的指针向量定义
        A * pa1=new A(1);
        A * pa2=new A(2);
        pca_vec.push_back(pa1);                                //添加类 A 的指针向量元素
        pca_vec.push_back(pa2);
        nSize=pca_vec.size();
        cout<<"通过数组输出:";
        for(int i=0; i<nSize; i++)                             //通过数组输出
        {
            cout<<pca_vec[i]->n<<"\t";
        }
        cout<<"\n通过引用输出:";
        for(int i=0; i<nSize; i++)                             //通过引用输出
        {
            A * &a=pca_vec.at(i);
            cout<<a->n<<"\t";
        }
        cout<<"\n通过迭代器输出:";
        vector<A *>::iterator pca_vec_iter=pca_vec.begin();    //通过迭代器输出
        while(pca_vec_iter!=pca_vec.end())
        {
            cout<<(* * pca_vec_iter).n<<"\t";
            pca_vec_iter++;
        }
        delete pa1;
        delete pa2;
        return 0;
    }
```

程序执行结果如下所示。

第一部分:类 A 的向量操作
      通过数组输出:1 2
      通过引用输出:1 2
      通过迭代器输出:1 2
第二部分:类 A 的指针向量操作
      通过数组输出:1 2
      通过引用输出:1 2
      通过迭代器输出:1 2

从源程序中,应主要分析出普通类、指针类变量所形成的向量中通过数组、引用、迭代器获得某向量元素写法上的不同。具体分析如下:

(1) 对于数组而言,vector<A> ca_vec 是类 A 定义变量的向量集,所以 ca_vec[i]就代表 A 定义的变量,ca_vec[i].n 表示对应 A 类对象的成员变量 n;vector<A*> pca_vec 是类 A 定义的指针变量的向量集,pca_vec[i]表示指向某对象的指针,pca_vec[i]—>n 表示某对象的成员变量 n。

(2) 对于引用而言,vector<A> ca_vec 是类 A 定义变量的向量集,A& a=ca_vec.at(i)获得的是对 A 对象的引用,所以 a.n 对应该对象的成员变量值;vector<A*> pca_vec 是类 A 定义的指针变量的向量集,A*&a=pca_vec.at(i)获得的是对象指针的引用,所以 a—>n 才表示真正的对象成员变量值。

(3) 对于迭代器而言,vector<A> ca_vec 是类 A 定义变量的向量集,是类 A 的对象集合,vector<A>::iterator ca_vec_iter 迭代器表示指向 A 对象的指针,所以 *ca_vec_iter 表示真正的 A 对象;vector<A*> pca_vec 是类 A 定义的指针变量的向量集,是 A* 指针的集合,vector<A*>::iterator pca_vec_iter 迭代器表示指向 A* 的指针,所以 *pca_vec_iter 表示 A* 指针的内容,**pca_vec_iter 表示真正的 A 对象。针对这种情况,可参考图 7.1 加以理解。

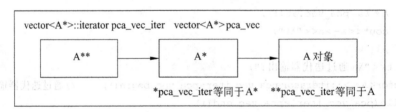

图 7.1 迭代器示意图

7.2.4 修改元素示例

通过对 7.2.3 节的学习,可知修改向量中的元素同样有三种主要方法:通过数组方式,通过引用方式及通过迭代器方式。

【例 7.4】 某整型向量先存入三个数,值为 1,2,3。先后修改第二个数为 5,10,20,并分别在屏幕上显示修改后的全部向量元素。

```
//文件名: e7_4.cpp
#include<vector>
#include<iostream>
using namespace std;

int main(int argc, char * argv[])
{
```

```
        vector<int>int_vec;
        int_vec.push_back(1);
        int_vec.push_back(2);
        int_vec.push_back(3);

        //通过数组修改
        cout<<"通过数组修改,第2元素为5: ";
        int_vec[1]=5;
        for(int i=0; i<int_vec.size(); i++)
        {
            cout<<int_vec[i]<<"\t";
        }
        cout<<endl;

        //通过引用修改
        cout<<"通过引用修改,第2元素为10: ";
        int &m=int_vec.at(1);
        m=10;
        for(int i=0; i<int_vec.size(); i++)
        {
            cout<<int_vec[i]<<"\t";
        }
        cout<<endl;

        //通过迭代器修改
        cout<<"通过迭代器修改,第2元素为20: ";
        vector<int>::iterator int_vec_iter=int_vec.begin()+1;
        *int_vec_iter=20;
        for(int i=0; i<int_vec.size(); i++)
        {
            cout<<int_vec[i]<<"\t";
        }
        return 0;
    }
```

程序执行结果如下所示。

```
通过数组修改,第2元素为5: 1    5    3
通过引用修改,第2元素为10: 1   10    3
通过迭代器修改,第2元素为20: 1   20    3
```

7.2.5 删除元素示例

删除元素主要是掌握 erase 函数的用法。

【例 7.5】 一个整型向量,初始化元素为 1~10。删除第 5 个元素,屏幕显示向量元素值;删除第 2~5 个元素,屏幕显示向量元素值。

```
//文件名: e7_5.cpp
#include<vector>
#include<iostream>
using namespace std;

int main(int argc, char * argv[])
{
    vector<int>int_vec;
    for(int i=1; i<=10; i++)
    {
        int_vec.push_back(i);
    }

    //删除第 5 个元素
    int_vec.erase(int_vec.begin() +4);
    cout<<"删除第 5 个元素后向量为: ";
    for(int i=0; i<int_vec.size(); i++)
    {
        cout<<int_vec[i]<<"\t";
    }

    //再删除第 2~5 个元素
    int_vec.erase(int_vec.begin() +1,int_vec.begin()+5);
    cout<<"再删除第 2~5 个元素后向量为: ";
    for(int i=0; i<int_vec.size(); i++)
    {
        cout<<int_vec[i]<<"\t";
    }
    cout<<endl;
    return 0;
}
```

程序执行结果如下所示。

删除第 5 个元素后向量为: 1 2 3 4 6 7 8 9 10
再删除第 2~5 个元素后向量为: 1 7 8 9 10

7.2.6 进一步理解 vector

通过图 7.2 来加以说明。

第 7 章 通用容器

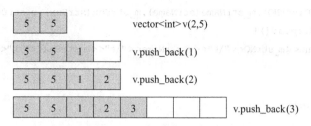

图 7.2　vector 内存示意图

当执行代码 vector<int>v(2,5)时,在内存里建立了 2 个整型元素空间,值是 5。当增加元素 1 时,原有空间由 2 个变为 4 个整型元素空间,并把元素 1 放入第 3 个整型空间,第 4 个空间作为预留空间。当增加元素 2 时,直接把值 2 放入第 4 个空间。当增加元素 3 时,由于原向量中没有预留空间,则内存空间由 4 个变为 8 个整型元素空间,并把值 3 放入第 5 个内存空间。

总之,增加新元素时,如果超过当前的容量,则容量会自动扩充 2 倍,如果两倍容量仍不足,就扩大至足够大的容量。本图是直接在原空间基础上画的新增空间,其实要复杂得多,包括重新配置、元素移动、释放原始空间的过程。因此对 vector 容器而言,当增加新的元素时,有可能很快完成(直接存在预留空间中),有可能稍慢(扩容后再放新元素);对修改元素值而言是较快的;对删除元素来说,若删除尾部元素较快,非尾部元素则稍慢,因为牵涉到删除后的元素移动。

7.2.7　综合操作示例

【例 7.6】 利用 vector 编一个学生信息[学号(它是关键字)、姓名、性别、出生日期]管理类,有添加函数、查询函数(依据学号查询)、显示函数(对查询结果完成显示)。并编制函数测试。

```
//文件名:e7_6.cpp
#include<iostream>
#include<vector>
#include<string>
using namespace std;
class Student                                           //学生类
{
public:
    string m_strNO;                                     //学号
    string m_strName;                                   //姓名
    string m_strSex;                                    //性别
    string m_strDate;                                   //出生日期
public:
    Student(string strNO,string strName,string strSex,string strDate):
```

```cpp
        m_strNO(strNO),m_strName(strName),m_strSex(strSex),m_strDate(strDate){}
    void Display(){
        cout<<m_strNO<<"\t"<<m_strName<<"\t"<<m_strSex<<"\t"<<m_strDate<<endl;
    }
};
class StudCollect
{
    vector<Student>m_vStud;
public:
    void Add(Student& s){
        m_vStud.push_back(s);
    }
    Student * Find(string strNO){
        bool bFind=false;
        int i=0;
        for(i=0; i<m_vStud.size(); i++)
        {
            Student& s=m_vStud.at(i);
            if(s.m_strNO ==strNO)
            {
                bFind=true;
                break;
            }
        }
        Student * s=NULL;
        if(bFind)
            s=&m_vStud.at(i);
        return s;
    }
};
int main()                                        //仿真测试程序
{
    Student s1("1001","zhangsan","boy","1985-10-10");
    Student s2("1002","lisi","boy","1984-6-10");
    Student s3("1003","wangwu","boy","1985-11-15");
    StudCollect s;
    s.Add(s1),s.Add(s2),s.Add(s3);
    Student * ps=s.Find("1002");
    if(ps)
        ps->Display();
    return 0;
}
```

主要理解点如下。

(1) 基本思想是要有一个基本信息类及该基本信息的集合类。对本题而言，基本信息类是学生类 Student，集合类是 StudCollect。

(2) Student 类中定义了 4 个基本的成员变量及屏幕显示函数 Display()。StudCollect 定义了一个成员变量 vector<Student>v，可看出本类是对学生集合对象的一个管理类，是实现集合管理类的根本所在。该类主要定义了 Add(Student& s) 函数及 Student * Find(string strNO) 函数。Add 函数非常好理解，就是向集合 v 中添加学生对象。对于根据学号查询函数 Find，可能有的学生会按如下代码写：

```
void Find(string strNO){
    for(int i=0; i<m_vStud.size(); i++)
    {
        Student& s=m_vStud.at(i);
        if(s.m_strNO==strNO)
        {
            cout<<s.m_strNO<<"\t"<<s.m_strName<<"\t"<<s.m_strSex<<"\t"<<s.m_strDate;
            break;
        }
    }
}
```

即把查询与显示功能都写在一起，从纯粹实现本题要求功能来讲是可以的，但从面向对象思考角度来看这样并不好，面向对象思想要求编程者不但要考虑目前要完成的功能，还要考虑到将来的维护和升级问题。若像上述写法，如果没有经过查询，就想在屏幕上显示某个学生对象的信息就不行了，但是原始写法是可行的，对下述代码成立：

```
Student s1("1005","zhaoliu","girl","1990-1-1");
s1.Display();
```

【例 7.7】 已知作者的基本信息有作者编号、姓名，书籍信息有书号、书名、出版社，一位作者可以出版许多书，如何用集合类体现出上述关系？并编制简单的测试类。

分析：根据题中要求程序必须能管理多个作者，每个作者又能管理其出版的多部书籍，因此应该是集合类嵌套集合类。所编制的功能类有书籍基本信息类 Book，作者基本信息类 Writer，作者集合信息类 WriterCollect，下述代码仅描述了"添加"功能。

```
//文件名：e7_7.cpp
#include<iostream>
#include<vector>
#include<string>
using namespace std;
class Book
{
    string m_strBookNO;                          //书号
```

```cpp
        string m_strBookName;                          //书名
        string m_strPublic;                            //出版社
    public:
        Book(string strNO,string strName,string strPublic):m_strBookNO(strNO),
            m_strBookName(strName),m_strPublic(strPublic)
        {}
    };
    class Writer
    {
        string m_strWriterNO;                          //作者编号
        string m_strWriterName;                        //作者姓名
        vector<Book>m_vecBook;                         //所拥有书的向量集合
    public:
        Writer(string strNO,string strName):m_strNO(strNO),m_strName(strName){}
        void AddBook(Book& book){                      //向该作者添加所著书籍
            m_vecBook.push_back(book);
        }
    };
    class WriterCollect
    {
        vector<Writer>m_vecWriter;
    public:
        void AddWriter(Writer& writer){                //添加新的作者
            m_vecWriter.push_back(writer);
        }
    };
    int main()
    {
        Writer w1("1001","zhangsan");                  //产生作者对象 w1
        Book b1("b001","aaa","public");                //产生两本书对象
        Book b2("b002","bbb","public");
        w1.AddBook(b1);                                //把这两本书加入作者对象 w1
        w1.AddBook(b2);

        Writer w2("1002","lisi");                      //产生作者对象 w2
        Book b3("b003","ccc","public");                //产生一本书对象
        w2.AddBook(b3);                                //把此书加入作者对象 w2

        WriterCollect collect;                         //产生作者集合类对象 collect
        collect.AddWriter(w1);                         //向 collect 对象中加入作者对象 w1
        collect.AddWriter(w2);                         //向 collect 对象中加入作者对象 w2
        return 0;
    }
```

主要理解点如下。

(1) Writer 作者类是一个基本信息类,但它包含了作者书籍的集合对象 vector<Book> m_vecBook,也就是说,基本信息类中也经常用到集合类。对初学集合类的学生而言,经常认为基本信息类中一定不包含集合类,其实这是一个误区。

(2) 集合类多层嵌套是常用的编程方法,若数据关系能划分成树型结构,且均满足一个根结点对应多个子结点,那么编程时就要考虑到集合类嵌套方法。

7.3 deque 容器

deque 容器为一个给定类型的元素进行线性处理,像向量一样,它能够快速地随机访问任一个元素,并且能够高效地插入和删除容器的尾部元素。但它又与 vector 不同,deque 支持高效插入和删除容器的头部元素,因此也叫做双端队列。deque 类常用的函数如下。

7.3.1 常用函数

(1) 构造函数。
- deque():创建一个空 deque。
- deque(int nSize):创建一个 deque,元素个数为 nSize。
- deque(int nSize, const T& t):创建一个 deque,元素个数为 nSize,且值均为 t。
- deque (const deque &):复制构造函数。

(2) 增加函数。
- void push_front(const T& x):双端队列头部增加一个元素 x。
- void push_back(const T& x):双端队列尾部增加一个元素 x。
- iterator insert(iterator it, const T& x):双端队列中某一元素前增加一个元素 x。
- void insert(iterator it, int n, const T& x):双端队列中某一元素前增加 n 个相同元素 x。
- void insert(iterator it,const_iterator first, const_iterator last):双端队列中某一元素前插入另一个相同类型向量的[first,last]间的数据。

(3) 删除函数。
- iterator erase(iterator it):删除双端队列中某一个元素。
- iterator erase(iterator first, iterator last):删除双端队列中[first,last]中元素。
- void pop_front():删除双端队列中最前一个元素。
- void pop_back():删除双端队列中最后一个元素。
- void clear():删除双端队列中所有元素。

(4) 遍历函数。
- reference at(int pos):返回 pos 位置元素的引用。

- reference front()：返回首元素的引用。
- reference back()：返回尾元素的引用。
- iterator begin()：返回向量头指针，指向第一个元素。
- iterator end()：返回向量尾指针，不包括最后一个元素，在其下面。
- reverse_iterator rbegin()：反向迭代器，最后一个元素的迭代指针。
- reverse_iterator rend()：反向迭代器，第一个元素前的迭代指针。

（5）判断函数。

bool empty() const：向量是否为空，若 true，则向量中无元素。

（6）大小函数。

- int size() const：返回向量中元素的个数。
- int max_size() const：返回最大可允许的双端队列元素数量值。

（7）其他函数。

- void swap(vector&)：交换两个同类型向量的数据。
- void assign(int n, const T& x)：设置容器大小为 n 个元素，每个元素值为 x。
- void assign(const_iterator first, const_iterator last)：容器中[first, last)中元素设置成当前双端队列元素。

deque 容器与 vector 容器的许多功能是相似的，因此下面示例主要演示不相同的部分。

7.3.2 基本操作示例

【例 7.8】 deque 中 push_front、pop_front 函数示例。

```
//文件名：e7_8.cpp
#include<iostream>
#include<deque>
using namespace std;
int main()
{
    deque<int>  d;
    d.push_back(10);
    d.push_back(20);
    d.push_back(30);
    cout<<"原始双端队列：";
    for(int i=0; i<d.size(); i++)
    {
        cout<<d.at(i)<<"\t";         //10 20 30
    }
    cout<<endl;
    d.push_front(5);
    d.push_front(3);
    d.push_front(1);
```

```
        cout<<"push_front(5,3,1)后:";
        for(int i=0; i<d.size(); i++)
        {
            cout<<d.at(i)<<"\t";                    //1 3 5 10 20 30
        }
        cout<<endl;
        d.pop_front();
        d.pop_front();
        cout<<"两次 pop_front 后:";
        for(int i=0; i<d.size(); i++)
        {
            cout<<d.at(i)<<"\t";                    //5 10 20 30
        }
        cout<<endl;
        return 0;
    }
```

deque 可以通过 push_front 直接在容器头增加元素，通过 pop_front 直接删除容器头元素，这一点是 vector 元素不具备的。

【例 7.9】 deque 与 vector 内存分配比较示例。

```
//文件名：e7_9.cpp
#include<iostream>
#include<vector>
#include<deque>
using namespace std;
int main()
{
    vector<int>v(2);
    v[0]=10;
    int * p=&v[0];
    cout<<"vector 第 1 个元素迭代指针 * p= " << * p<<endl;         //10
    v.push_back(20);
    cout<<"vector 容量变化后原 vector 第 1 个元素迭代指针 * p="<< * p<<endl;
                                                              //数不确定

    deque<int>d(2);
    d[0]=10;
    int * q=&d[0];
    cout<<"deque 第 1 个元素迭代指针 * q= " << * q<<endl;          //10
    d.push_back(20);
    cout<<"deque 容量变化后第 1 个元素迭代指针 * q= " << * q<<endl;  //10
    return 0;
}
```

执行结果为：

vector 第 1 个元素迭代指针 * p=10
vector 容量变化后原 vector 第 1 个元素迭代指针 * p=-572662307
deque 第 1 个元素迭代指针 * q=10
deque 容量变化后第 1 个元素迭代指针 * q=10

该段程序功能是：deque、vector 初始化大小为 2，第 1 个元素都为 10，当通过 push_back 函数分别给两容器增加一个元素后，从结果发现原先保持的指针元素值对 vector 容器前后发生了变化，而对 deque 容器前后没有发生变化。图 7.3 演示了该示例中 vector、deque 内存变化的过程。

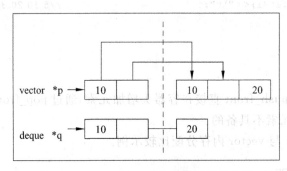

图 7.3　vector、deque 内存变化假想示意图

push_back(20)前如虚线左侧。push_back(20)后，对 vector 而言：由于当前有效总空间是 2，当增加元素 20 后，必须有 3 个空间，这样 vector 就重新分配空间，原先的两个元素复制到新空间中，原先空间释放，因此 * p 在 push_back 前后是变化的；对 deque 而言：当前有效总空间是 2，当增加元素 20 后，原有空间仍然保持，再建立一个新的内存块，把 20 放入其中，因此 * p 在 push_back 前后是不变化的。

虽然图 7.3 在细节上可能与真实的内存分配有很大差距，但在关键问题上是一致的：在建立 vector 容器时，一般来说伴随着建立空间→填充数据→重建更大空间→复制原空间数据→删除原空间→添加新数据，如此反复，保证 vector 始终是一块独立的连续内存空间；在建立 deque 容器时，一般伴随着建立空间→填充数据→建立新空间→填充新数据，如此反复，没有原空间数据的复制和删除过程，是由多个分段连续的内存空间组成的。

7.3.3　综合操作示例

【例 7.10】 以 deque 为基础，编一个先进先出的队列类。

先进先出队列类应具有下列函数：

（1）队头插入函数 push；
（2）队尾删除函数 pop；
（3）队列大小函数 size；

(4) 队列空判定函数 empty；
(5) 获得队头元素函数 front；
(6) 获得队尾元素函数 back。

通常编制队列的方法是：定义数据结构，对具体的函数要一步步完成具体的代码，也就是说要具体完成上述的 6 个函数。但是，仔细分析一下，队列操作仅是对队头完成删除操作，对队尾完成插入操作，它所需要的功能完全在 deque 容器提供功能的范围之内。那么很自然想到从 deque 中提取需要的函数，这样具体的代码就不用重新编制了。

编制的代码如下：

```cpp
//文件名：e7_10.cpp
#include<iostream>
#include<deque>
using namespace std;
template <class T>
class MyQueue
{
    deque<T>d;
public:
    void push(const T& t)                   //添加队尾元素
    {
        d.push_back(t);
    }
    void pop()                              //删除队头元素
    {
        d.pop_front();
    }
    int size()                              //获得队列大小
    {
        return d.size();
    }
    bool empty()                            //队列空否
    {
        return d.empty();
    }
    T& front()                              //获得队头元素
    {
        return *d.begin();
    }
    T& back()                               //获得队尾元素
    {
        return *(--d.end());
    }
```

```
        }
        void display()
        {
            for(int i=0; i<d.size(); i++)
            {
                cout<<d.at(i)<<"\t";
            }
        }
    };
    int main()                                          //测试函数
    {
        MyQueue<int>myqueue;
        for(int i=1; i<=5; i++)                         //初始化队列:1 2 3 4 5
        {
            myqueue.push(i);
        }
        cout<<"原队列:";
        myqueue.display(); cout<<endl;                  //原队列显示:1 2 3 4 5
        myqueue.pop();                                  //删除队头元素 1
        cout<<"删除队头元素后:";
        myqueue.display(); cout<<endl;
        myqueue.push(6);                                //插入队尾元素 6
        cout<<"插入队尾元素 6 后:";
        myqueue.display(); cout<<endl;
        cout <<"当前队头元素:";
        cout<<myqueue.front()<<endl;                    //获得队头元素
        cout <<"当前队尾元素:";
        cout<<myqueue.back()<<endl;                     //获得队尾元素
        return 0;
    }
```

本例中巧妙地重新封装了 deque 类,通过接口转换完成了队列类功能的实现,这其实是容器适配器思想,后面还要讲到。但在同学们脑海中要加深一下这个思想:不要总是被动地应用 STL 类,或许巧妙组合封装 STL 固有知识就能开发出更强大的功能代码来。

7.4 list 容器

相对于 vector 的连续线性空间,list 是一个双向链表,它有一个重要性质:插入操作和删除操作都不会造成原有的 list 迭代器失效,每次插入或删除一个元素就配置或释放一个元素空间。也就是说,对于任何位置的元素插入或删除,list 永远是常数时间。

7.4.1 常用函数

(1) 构造函数。
- list<Elem>c：创建一个空的 list。
- list<Elem>c1(c2)：复制另一个同类型元素的 list。
- list<Elem>c(n)：创建 n 个元素的 list,每个元素值由默认构造函数确定。
- list<Elem>c(n,elem)：创建 n 个元素的 list,每个元素值为 elem。
- list<Elem>c(begin,end)：由迭代器创建 list,迭代区间为[begin,end)。

(2) 大小、判断空函数。
- int size() const：返回容器元素个数。
- bool empty() const：判断容器是否空,若返回 true,表明容器已空。

(3) 增加、删除函数。
- void push_back(const T& x)：list 容器尾元素后增加一个元素 x。
- void push_front(const T& x)：list 容器首元素前增加一个元素 x。
- void pop_back()：删除容器尾元素,当且仅当容器不为空。
- void pop_front()：删除容器首元素,当且仅当容器不为空。
- void remove(const T& x)：删除容器中所有元素值等于 x 的元素。
- void clear()：删除容器中所有元素。
- iterator insert(iterator it, const T& x=T())：在迭代器指针 it 前插入元素 x,返回 x 迭代器指针。
- void insert(iterator it, size_type n, const T& x)：在迭代器指针 it 前插入 n 个相同元素 x。
- void insert(iterator it, const_iterator first, const_iterator last)：把[first, last)间的元素插入迭代指针 it 前。
- iterator erase(iterator it)：删除迭代器指针 it 对应的元素。
- iterator erase(iterator first, iterator last)：删除迭代器指针[first, last)间的元素。

(4) 遍历函数。
- iterator begin()：返回首元素的迭代器指针。
- iterator end()：返回尾元素后的迭代器指针,而不是尾元素的迭代器指针。
- reverse_iterator rbegin()：返回尾元素的逆向迭代器指针,用于逆向遍历容器。
- reverse_iterator rend()：返回首元素前的逆向迭代器指针,用于逆向遍历容器。
- reference front()：返回首元素的引用。
- reference back()：返回尾元素的引用。

(5) 操作函数。
- void sort()：容器内所有元素排序,默认是升序。

- template ＜class Pred＞void sort(Pred pr)：容器内所有元素根据预判定函数 pr 排序。
- void swap(list& str)：两 list 容器交换功能。
- void unique()：容器内相邻元素若有重复的，则仅保留一个。
- void splice(iterator it，list& x)：队列合并函数，队列 x 所有元素插入迭代指针 it 前，x 变成空队列。
- void splice(iterator it，list& x，iterator first)：队列 x 中移走[first,end)间元素插入迭代指针 it 前。
- void splice(iterator it，list& x，iterator first，iterator last)：x 中移走[first,last)间元素插入迭代指针 it 前。
- void reverse()：反转容器中元素顺序。

7.4.2 基本操作示例

【例 7.11】 push_back、push_front、pop_front、pop_back、begin、rbegin 示例。

```cpp
//文件名：e7_11.cpp
#include<iostream>
#include<string>
#include<list>
using namespace std;
typedef list<string> LISTSTR;

int main()
{
    LISTSTR test;

    test.push_back("back");                //back
    test.push_front("middle");             //middle back
    test.push_front("front");              //front middle back

    cout<<test.front()<<endl;              //front
    cout<<*test.begin()<<endl;             //front

    cout<<test.back()<<endl;               //back
    cout<<*(test.rbegin())<<endl;          //back

    test.pop_front();                      //middle back
    test.pop_back();                       //middle

    cout<<test.front()<<endl;              //middle
    return 0;
}
```

(1) 可看出对 list 首尾元素的增加和删除都是非常容易的。

(2) test.front()相当于 string& s=test.front(),返回了首元素的引用;test.begin()相当于 list<string>::iterator it=test.begin(),返回了首元素的迭代器指针。因此 test.front()与 * test.begin()的结果是一致的。

【例 7.12】 遍历示例。

```
//文件名：e7_12.cpp
#include<iostream>
#include<list>
using namespace std;
typedef list<int>LISTINT;

int main()
{
    LISTINT test;
    for(int i=0; i<5; i++)
    {
        test.push_back(i+1);
    }
    //show
    LISTINT::iterator it=test.begin();
    for(; it!=test.end(); it++)
    {
        cout<< * it<<"\t";                      //1 2 3 4 5
    }
    cout<<endl;
    //reverse_show
    LISTINT::reverse_iterator rit=test.rbegin();
    for(; rit!=test.rend(); rit++)
    {
        cout<< * rit<<"\t";                     //5 4 3 2 1
    }
    cout<<endl;
    return 0;
}
```

注意：正向迭代器与逆向迭代器表示形式是不一样的,前者是 iterator,后者是 reverse_iterator。另外,逆向显示并不改变元素在容器中的位置,只是显示逆向了。

【例 7.13】 sort、merge、splice 示例。

```
//文件名：e7_13.cpp
#include<iostream>
#include<list>
```

```cpp
using namespace std;
typedef list<int>LISTINT;

int main()
{
    LISTINT t1;
    t1.push_back(1);
    t1.push_back(5);
    t1.push_back(3);
    t1.push_back(10);                                       //1 5 3 10

    LISTINT t2;
    t2.push_back(2);
    t2.push_back(8);
    t2.push_back(6);
    t2.push_back(9);                                        //2 8 6 9

    t1.sort();                                              //1 3 5 10
    t2.sort();                                              //2 6 8 9

    //t1.splice(t1.begin(), t2);                            //2 6 8 9 1 3 5 10
    t1.merge(t2);                                           //1 2 3 5 6 8 9 10

    for(LISTINT::iterator it=t1.begin(); it!=t1.end(); it++)
    {
        cout<< * it<<"\t";
    }
    cout<<endl;
    cout<<t1.size()<<"\t"<<t2.size()<<endl;                 //8 0
    return 0;
}
```

(1) 两个链表 merge 合并前，一般都已经按升序排好序，合并后的链表元素仍然是升序序列。

(2) merge 操作是数据移动操作，不是复制操作，因此 t1.merge(t2)表示把 t2 中所有元素依次移动并插入到源链表 t1 的适当位置，t1 增加了多少个元素，t2 就减少了多少个元素。

(3) 若用 t1.splice(t1.begin()，t2)代替程序中的 t1.merge(t2)，其余不变，就能看出 splice 的特点。splice 完成的是拼接功能，也是数据移动操作，不是复制操作。t1.splice(t1.begin()，t2)表明把 t2 中所有元素整体地移动到原始链表 t1 的首元素前，t1 增加了多少个元素，t2 就减少了多少个元素。如题中所述 t1、t2 排序后，t1={1,3,5,10},t2={2,6,8,9}。t1.splice(t1.begin()，t2)后，t1={2,6,8,9,1,3,5,10},t2={}；t1.merge(t2)后，

t1={1,2,3,5,6,8,9,10}，t2={}。

7.4.3 综合操作示例

【例 7.14】 两个文本文件中包含某中学的高考成绩,包含准考证号、姓名、所考大学名、总成绩信息,准考证号是关键字。但可能由于一些原因,造成两个文件中有重复的记录,现要求把两个文件内容合并在一起,去掉重复记录,并按准考证号升序排列。

分析：
(1) 把两个文本文件数据映射成两个 list 容器中的元素。
(2) 对两个 list 容器分别按准考证号进行升序排序,利用函数是 sort。
(3) 合并两个已排好序的 list 容器元素,利用函数是 merge。
(4) 利用 unique 函数去掉准考证号重复的记录,仅保留一个即可。

```cpp
//文件名：e7_14.cpp
#include<iostream>
#include<list>
#include<string>
using namespace std;
class Student
{
private:
    string m_strNO;                            //准考证号
    string m_strName;                          //姓名
    string m_strUniversity;                    //大学
    int    m_nTotal;                           //高考成绩
public:
    Student(string strNO,string strName,string strUniversity,int nTotal):
            m_strNO(strNO),m_strName(strName),
            m_strUniversity(strUniversity),m_nTotal(nTotal)
            {
            }
    bool operator <(Student& s)
    {
        return m_strNO<s.GetNO();
    }
    bool operator==(Student& s)
    {
        return m_strNO ==s.GetNO();
    }
    string GetNO() {return m_strNO;}
    string GetName() {return m_strName;}
```

```cpp
        string GetUniversity() {return m_strUniversity;}
        int    GetTotal(){return m_nTotal;}
};

ostream& operator<<(ostream& os, Student& s)
{
    os<<s.GetNO()<<"\t"<<s.GetName()<<"\t"<<s.GetUniversity()<<"\t" <<s.GetTotal();
    return os;
}

typedef list<Student>  LISTSTUD;

class StudManage
{
private:
    LISTSTUD m_stlList;
public:
    bool Add(const Student& s)
    {
        m_stlList.push_back(s);
        return true;
    }
    bool Merge(StudManage& stud)
    {
        m_stlList.sort();
        stud.m_stlList.sort();
        m_stlList.merge(stud.m_stlList);
        m_stlList.unique();
        return true;
    }
    void Show()
    {
        for(LISTSTUD::iterator it=m_stlList.begin(); it!=m_stlList.end(); it++)
        {
            cout<< * it<<endl;
        }
    }
};

int main()
{
    StudManage sm1;
    StudManage sm2;
```

```
        Student s1("1001", "zhangsan","tsinghua", 670);
        Student s2("1002", "lisi","beida", 660);
        Student s3("1003", "wangwu","fudan", 650);
        Student s4("1004", "zhaoliu","nankai", 640);
        Student s5("1005", "zhouqi","tongji", 630);

        sm1.Add(s1);
        sm1.Add(s2);
        sm1.Add(s5);

        sm2.Add(s5);
        sm2.Add(s4);
        sm2.Add(s3);
        sm2.Add(s1);

        sm1.Merge(sm2);
        sm1.Show();
        return 0;
    }
```

(1) 由于本示例主要对 list 进行操作,把两个文本文件数据映射成两个 list 容器中的元素采取了仿真,直接写在了主程序 main 中:"第 1 个文件"有 3 个学生对象 s1,s2,s5,保存在容器 sm1 中;"第 2 个文件"有 4 个学生对象 s5,s4,s3,s1,保存在容器 sm2 中。两个文件的重复对象是学生对象 s1。

(2) 有一个基本类 Student,一个集合维护类 StudManage,一个重载的 ostream& operator<<(ostream& os, Student& s)函数,这是由于要利用 cout 输出学生对象的缘故。

(3) StudManage 类 Merge 成员函数封装了两个 list 容器的排序,合并及去掉重复元素功能。

(4) 由于对 list 容器均是按学号进行排序、合并的,因此要重载基本类 Student 中的 operator<操作符。又由于要按学号是否相同去掉重复记录,因此必须重载基本类 Student 中的 operator==操作符。

【例 7.15】 编制多项式加法的类,例如若 $p1=2x^6+3x^4+5x^2+6$, $p2=2x^5-3x^4+5x^2+8$,则 $p3=p1+p2=2x^6+2x^5+10x^2+14$。

分析:

(1) 多项式用 list 容器表示,容器内的每个元素代表多项式中的每一项。设置多项式 p1,p2,并使结果 p3 为空。

(2) 利用 sort 系统函数,使多项式 p1,p2 中每一项按系数升序排列。

(3) 利用 iterator 技术遍历 p1,p2。当 p1,p2 没有检测完各自的链表时,比较当前结点的指数域。

(4) 指数相等,对应项系数相加。若相加结果不为 0,则结果加入 p3,否则不加入 p3。

p1、p2各自的iterator迭代指针都进1。

(5) 指数不等，小系数者加入p3，相应iterator迭代指针进1。

(6) 当p1、p2的iterator迭代指针已有一个检测完毕，把另一个链表的剩余部分加入p3即可，采用函数是系统函数splice。

程序代码如下所示。

```cpp
//文件名：e7_15.cpp
#include<iostream>
#include<list>
#include<string>
using namespace std;

class Term                                      //多项式项描述类
{
private:
    int coef;                                   //系数
    int exp;                                    //指数
public:
    Term(int coef, int exp):coef(coef),exp(exp)
    {
    }
    int GetCoef(){return coef;}
    int GetExp(){return exp;}
    bool operator< (Term& t)
    {
        return exp <t.GetExp();
    }
};

typedef list<Term>LISTTERM;

ostream& operator<<(ostream& os, Term& t)
{
    os<<t.GetCoef()<<"x"<<t.GetExp();
    return os;
}

class Polynomial                                //多项式操作类
{
private:
    LISTTERM m_stlTermList;
public:
```

```cpp
bool AddTerm(const Term& t)                    //向多项式链表中添加每一项
{
    m_stlTermList.push_back(t);
    return true;
}

Polynomial AddPolynomial(Polynomial& obj)      //计算两个多项式的和
{
    Polynomial result;

    m_stlTermList.sort();                      //当前链表排序
    obj.m_stlTermList.sort();                  //被加链表排序

    LISTTERM::iterator src=m_stlTermList.begin();
    LISTTERM::iterator des=obj.m_stlTermList.begin();

    int coef=0;
    int exp =0;

    while((src!=m_stlTermList.end()) && (des!=obj.m_stlTermList.end()))
    {
        int nCurExp1=(*src).GetExp();
        int nCurExp2=(*des).GetExp();

        if(nCurExp1 ==nCurExp2)
        {
            coef=(*src).GetCoef()+(*des).GetCoef();
            exp=(*src).GetExp();
            src++;
            des++;
        }
        if(nCurExp1 <nCurExp2)
        {
            coef=(*src).GetCoef();
            exp=(*src).GetExp();;

            src++;
        }
        if(nCurExp1 >  nCurExp2)
        {
            coef=(*des).GetCoef();
            exp=(*des).GetExp();
```

```
            des++;
        }
        if(coef!=0)                         //如果系数不为零,则保存至结果
        {
            Term t(coef, exp);
            result.AddTerm(t);
        }
    }

    //加多余部分的链表
    if(src!=m_stlTermList.end())
    {
        LISTTERM temp(src, m_stlTermList.end());
        result.m_stlTermList.splice(result.m_stlTermList.end(), temp);
    }
    if(des!=m_stlTermList.end())
    {
        LISTTERM temp(des, obj.m_stlTermList.end());
        result.m_stlTermList.splice(result.m_stlTermList.end(), temp);
    }

    return result;
}
//简单显示多项式函数,可改进
void Show()
{
    LISTTERM::iterator it=m_stlTermList.begin();
    while(it!=m_stlTermList.end())
    {
        cout<< * it<<"+";

        it++;
    }
}
};

int main()
{
    Polynomial p1;
    Polynomial p2;

    Term t1(2, 6);
    Term t2(3, 4);
    Term t3(5, 2);
```

```
        Term t4(6, 0);
        p1.AddTerm(t1);
        p1.AddTerm(t2);
        p1.AddTerm(t3);
        p1.AddTerm(t4);

        Term t5(2, 5);
        Term t6(-3, 4);
        Term t7(5, 2);
        Term t8(8, 0);
        p2.AddTerm(t5);
        p2.AddTerm(t6);
        p2.AddTerm(t7);
        p2.AddTerm(t8);

        Polynomial p3=p1.AddPolynomial(p2);
        p3.Show();
        return 0;
    }
```

(1) Term 类中重载 operator＜运算符,是因为 Polynomial 中对 list＜Term＞容器 m_stlTermList按指数升序排序的需要。

(2) 重载函数 ostream& operator<<(ostream& os, Term& t),是因为要标准输出一个 Term 多项式项。

(3) 当执行 main()函数 p3=p1.AddPolynomial(p2)时,必须保证不能修改原始多项式 p1,p2。因此在 AddPolynomial 函数中要格外小心,特别是按算法操作多余多项式项时,即利用系统函数 splice 要注意,必须利用构造函数产生一个原始链表的备份,再执行 splice 才可以。

7.5 队列和堆栈

队列和栈是常用和重要的数据结构。队列只允许在表的一端插入,在另一端删除,允许插入的一端叫做队尾,允许删除的一端叫做队头,是一种先进先出线性表。栈只允许在表的一端进行插入和删除操作,是一种后进先出的线性表。

7.5.1 常用函数

(1) 构造函数。
- queue(class T, class Container＝deque＜T＞):创建元素类型为 T 的空队列,默认容器是 deque。
- stack(class T, class Container＝deque＜T＞):创建元素类型为 T 的空堆栈,默认

容器是 deque。

(2) 操作函数。

队列和堆栈共有函数：
- bool empty()：如果队列(堆栈)为空返回 true，否则返回 false。
- int size()：返回队列(堆栈)中元素数量。
- void push(const T& t)：把 t 元素压入队尾(栈顶)。
- void pop()：当队列(栈)非空情况下，删除队头(栈顶)元素。

队列独有函数：
- T& front()：当队列非空情况下，返回队头元素引用。
- T& back()：当队列非空情况下，返回队尾元素引用。

堆栈独有函数：
T& top()：当栈非空情况下，返回栈顶元素的应用。

7.5.2 容器配接器

为了清楚容器配接器概念，先看一段标准模板库中 queue 类的代码。

```
template<class T, class Container=deque<T>>
class queue
{
    protected:
        Container c;
    public:
        void push(T &t) {c.push_back(t);}
        ...
};
```

可以看出如下特点。

(1) 成员变量是标准模板库基础容器类 Container 对应的变量 c，队列的各个元素存在于容器 c 中。从上述 push 函数中可以得出：它是通过调用 Container 类中基础函数 push_back 完成队列的插入操作的。具体来说，当默认容器是 deque，就调用 deque 类中的 push_back 函数；当传入容器参数是 list，就调用 list 中的 deque 函数。因此 queue 中各操作函数只是起一个配接作用，几乎没有自己独有的功能。广而言之，queue 类是对基础容器类 Container 的再封装，不是重新定义。表面上操作的是队列类 queue 的各个函数，其实操作的是转接后 Container 类中的函数，queue 只是起一个中介作用，这就是容器配接器的概念。

(2) 能成为 queue 基本容器类 Container 的条件是它应当支持 size、empty、push_back、pop_front、front、back 方法，可对数据的两端分别进行插入、删除操作，而 deque、list 都具有这些函数，所以它们可成为 queue 的基本容器类 Container；能成为 stack 基本容器类 Container 的条件是它应当支持 size、empty、push_back、pop_back、back 方法，可对数据的一端进行插入、删除操作，而 deque、list、vector 都具有这些函数，所以它们可成为 stack 的基本容器类 Container。注意：vector 不能作为 queue 的基本容器类，因为 vector 没有 pop_front 方法。

7.5.3 基本操作示例

【例 7.16】 stack 基本函数操作示例。

```
//文件名：e7_16.cpp
#include<iostream>
#include<string>
#include<vector>
#include<list>
#include<stack>
using namespace std;

void PrintStack(stack<int, vector<int>>obj)        //堆栈遍历函数 1
{
    while(!obj.empty())
    {
        cout<<obj.top()<<"\t";
        obj.pop();
    }
}

void PrintStack(stack<string, list<string>>obj)    //堆栈遍历函数 2
{
    while(!obj.empty())
    {
        cout<<obj.top()<<"\t";
        obj.pop();
    }
}

template <class T, class Container>
void PrintStack(stack<T, Container>obj)            //堆栈遍历模板函数
{
    while(!obj.empty())
    {
        cout<<obj.top()<<"\t";
        obj.pop();
    }
```

```cpp
int main()
{
    stack<int, vector<int>>s;                       //整型堆栈
    for(int i=0; i<4; i++)
    {
        s.push(i+1);
    }
    PrintStack(s);                                   //4 3 2 1
    cout<<endl;

    string str="a";                                  //字符串堆栈
    stack<string, list<string>>t;
    for(int i=0; i<4; i++)
    {
        t.push(str);
        str+="a";
    }
    PrintStack(t);                                   //aaaa aaa aa a
    cout<<endl;

    stack<float, deque<float>>u;                     //浮点堆栈
    for(int i=0; i<4; i++)
    {
        u.push(i+1);
    }
    PrintStack(u);                                   //4 3 2 1
    return 0;
}
```

（1）main 函数中分别利用基本容器类 vector、list、deque 形成了三个堆栈，再调用所需的 PrintStack 函数完成堆栈元素的显示。PrintStack 函数中首先通过 empty 函数判断堆栈是否为空，若不空，则通过 top 函数获得栈顶元素的引用并显示，然后通过 pop 函数删除栈顶元素。如此反复，直至堆栈为空。

（2）显然，main 函数中 PrintStack(s) 调用的是第一个函数 PrintStack(stack＜int，vector＜int＞＞obj)，PrintStack(t) 调用的是第二个函数 PrintStack(stack＜string，list＜string＞＞obj)，PrintStack(u) 调用的是第三个函数 PrintStack(stack＜T, Container＞obj)。前两个函数参数类型都是具体的，第三个是堆栈的模板遍历函数，参数是不确定的，通用性更广。其实可以把前两个函数去掉，其他不变，显示结果仍然一样，在这里写出来是让同学们比较一下。

(3) PrintStack 函数参数是复制类型，不能写成引用类型。例如写成 PrintStack(stack<int，vector<int>>&obj)，则当该函数执行时实参和行参都指向相同的实体，函数执行完后堆栈变成空的了，相当于原先的堆栈实体内容变了，因此是错误的。

【例 7.17】 queue 基本函数操作示例。

```
//文件名：e7_17.cpp
#include<iostream>
#include<string>
#include<list>
#include<deque>
#include<queue>
using namespace std;
template<class T, class Container>              //队列遍历模板函数
void PrintQueue(queue<T, Container>obj)
{
    while(! obj.empty())
    {
        cout<<obj.front()<<"\t";
        obj.pop();
    }
}

int main()
{
    string str="a";                              //字符串队列
    queue<string, deque<string>>t;
    for(int i=0; i<4; i++)
    {
        t.push(str);
        str+="a";
    }
    PrintQueue(t);                               //a aa aaa aaaa
    cout<<endl;
    queue<float, list<float>>u;                  //浮点队列
    for(int i=0; i<4; i++)
    {
        u.push(i+1);
    }
    PrintQueue(u);                               //1 2 3 4
    return 0;
}
```

本例代码演示了 list、deque 形成的队列及其遍历功能。注意，queue 容器不能以 vector 作为基本容器类，这在前文中已论述过。

7.5.4 综合操作示例

【**例 7.18**】 编一个固定大小的堆栈类。

分析：标准模板库中 stack 中元素个数没有限制，与题意不符，那么如何既能用上 stack 固有功能，又能加上特有的限制呢？其实在这句话中已经体现出了具体思路：(1) 从 stack 派生出自己定义的堆栈类 mystack。这样，mystack 类就继承了 stack 的固有特性。(2) 在 mystack 类中加入特有的功能，题意中要求限制大小，那么一方面要定义一个成员变量 m_nMaxSize，通过构造函数传入堆栈大小，并赋给 m_nMaxSize；另一方面要重载 push 函数，如果当前堆栈元素个数小于 m_nMaxSize，则把新元素压入堆栈。

```cpp
//文件名：e7_18.cpp
#include<iostream>
#include<deque>
#include<stack>
using namespace std;

template<class T, class Container=deque<T>>
class mystack : public stack<T, Container>
{
private:
    int m_nMaxSize;                                    //堆栈大小
public:
    mystack(int maxsize)
    {
        m_nMaxSize=maxsize;
    }

    void push(const T &t)                              //重载 push 函数
    {
        if(stack<T,Container>::size()<m_nMaxSize)      //如果堆栈元素个数小于 m_nMaxSize
        {
            stack<T, Container>::push(t);              //则压入堆栈
        }
        else                                           //否则堆栈已满
        {
            cout<<"stack is fill."<<"the term "<<t<<" is not pushed"<<endl;
        }
    }
```

```
};
int main()
{
    mystack<int,deque<int>>obj(2);        //设置堆栈大小为 2
    obj.push(1);                          //可以入栈,size=1
    obj.push(2);                          //可以入栈,size=2
    obj.push(3);                          //栈已满,不能入栈
    return 0;
}
```

同学们可仿照该例,编制一个固定大小的队列类。

【例 7.19】 假设表达式中允许包含两种括号:圆括号和方括号,其嵌套的顺序随意,如[()]、([])()等是正确的表示。而([)]、(()]等是非法的表达式。编制相应的类。

分析:以括号序列[([])]为例说明。当计算机接收到第 1 个括号"["后,它盼望与其匹配的括号"]"出现;接收到的第 2 个括号"("不与第 1 个括号"["匹配,暂时放在一边;接收到的第 3 个括号"["不与第 2 个括号"("匹配,暂时放在一边;接收到的第 4 个括号"]"正好与第 3 个括号"["匹配;接收到的第 5 个括号")"正好与第 2 个括号"("相匹配;接收到的第 6 个括号"]"正好与第 1 个括号"["相匹配。可以看出匹配的特点是后读进计算机的左括号却是最先找到匹配元素的,这正是栈的特点,因此得出算法如下。

(1) 初始化堆栈为空,合法表达式标识 flag=true(假设表达式都是正确的)。

(2) 循环读括号表达式的每一个元素。

(3) 若是左括号,则把它压入栈中。

(4) 若是右括号,则读栈内容,有两种情况:若堆栈空,说明表达式不正确,如")[]",则置 flag=false,跳转至步骤(6)。若堆栈不空,则读栈顶元素值,若该括号值正好与右括号匹配,则把栈顶元素弹出;若该括号值不与右括号匹配,则是非法表达式,如"(]",则置 flag=false,跳转至步骤(6)。

(5) 若堆栈为空,则说明所有的左括号都找到了与之匹配的右括号,是合法的表达式;若堆栈不为空,说明所有的右括号都有与之匹配的左括号,但还有多余的左括号,如"(([]",则置 flag=false。

(6) 把标志 flag 返回调用程序。

总之,该算法的核心有两点:

(1) 当读到右括号时,则读栈顶元素,若正是匹配的左括号,则删除栈顶元素。

(2) 栈初始时是空的,若括号表达式是完全正确的,则经过一系列的入栈、出栈后,栈最终仍然是空的。

类代码及测试程序如下所示。

```
//文件名:e7_19.cpp
#include<iostream>
#include<string>
#include<stack>
```

```cpp
using namespace std;

class CExpress
{
private:
    string m_strExpress;
public:
    CExpress(string strExpress):m_strExpress(strExpress)
    { }
    bool IsValid()
    {
        bool flag=true;                        //假设表达式都是正确的
        char ch=0;
        char chstack=0;
        stack<char>sta_char;                   //堆栈初始化为空
        int size=m_strExpress.length();        //获取字符串长度

        for(int i=0; i<size; i++)
        {
            ch=m_strExpress.at(i);             //利用 at 函数读字符串的每个字符
            if(ch=='('||ch=='[')               //如果左括号,则压入堆栈
            {
                sta_char.push(ch);
            }

            if(ch==')'||ch==']')               //如果右括号,有下述情况
            {
                if(sta_char.empty())           //1.堆栈为空,表明表达式有多余右括号
                {
                    flag=false;                //置表达式错误标识
                    break;
                }
                else
                {
                    chstack=sta_char.top();    //2.堆栈不为空,读栈顶元素
                    if(chstack=='('&& ch==')') //匹配情况 1
                    {
                        sta_char.pop();        //删除栈顶元素
                    }
                    else if(chstack=='['&& ch==']')//匹配情况 2
                    {
                        sta_char.pop();        //删除栈顶元素
                    }
                    else                       //错误情况,置错误标识
                    {
                        flag=false;
```

```
                break;
            }
         }
       }                                      //else
     }                                        //if(ch==')'||ch==']')
                                              //for(int i=0; i<size; i++)
     if(flag==true && !sta_char.empty())      //栈中有多余左括号,置错误标识
     {
         flag=false;
     }
     return flag;
   }
};
int main()
{
    CExpress c1("()[]");
    CExpress c2("([]");
    CExpress c3("()[");
    CExpress c4("([]())");

    c1.IsValid()==true?cout<<"c1 is right"<<endl: cout<<"c1 is wrong"<<endl;
    c2.IsValid()==true?cout<<"c2 is right"<<endl: cout<<"c2 is wrong"<<endl;
    c3.IsValid()==true?cout<<"c3 is right"<<endl: cout<<"c3 is wrong"<<endl;
    c4.IsValid()==true?cout<<"c4 is right"<<endl: cout<<"c4 is wrong"<<endl;
    return 0;
}
```

7.6 优先队列

优先队列即 priority_queue 类,带优先权的队列,优先权高的元素优先出队。与普通队列相比,共同点都是对队头做删除操作,队尾做插入操作,但不一定遵循先进先出原则,也可能后进先出。priority_queue 是一个基于某个基本序列容器进行构建的适配器,默认的序列容器是 vector。

7.6.1 常用函数

(1) 构造函数。
- **priority_queue**(const Pred& pr=Pred(), const allocator_type& al=allocator_type()):创建元素类型为 T 的空优先队列,Pred 是二元比较函数,默认是 less<T>。
- **priority_queue**(const value_type * first, const value_type * last, const Pred& pr=Pred(), const allocator_type& al=allocator_type()):以迭代器[first, last]指向元

素,创建元素类型为 T 的优先队列,Pred 是二元比较函数,默认是 less<T>。
(2) 操作函数。
- bool empty():如果优先队列为空返回 true,否则返回 false。
- int size():返回优先队列中元素数量。
- void push(const T& t):把 t 元素压入优先队列。
- void pop():优先队列非空情况下,删除优先级最高元素。
- T& top():优先队列非空情况下,返回优先级最高元素的引用。

7.6.2 基本操作示例

【例 7.20】 演示整型序列进出 priority_queue。

```
//文件名:e7_20.cpp
#include<iostream>
#include<queue>
#include<algorithm>
#include<iterator>
using namespace std;
int main()
{
    int a[]={1,2,3,4,5,6,7,8,9,10};
    priority_queue<int>pr(a, a+9);       //通过构造函数将 a[0]~a[8]送入优先队列
    pr.push(a[9]);                        //通过函数将 a[9]送入优先队列

    cout<<"进队顺序:";
    copy(a, a+10, ostream_iterator<int>(cout, "\t"));
    cout<<endl;

    cout<<"出队顺序:";
    while(!pr.empty())
    {
        cout<<pr.top()<<"\t";             //获得优先队列队头元素
        pr.pop();                          //删除队头元素
    }
    return 0;
}
```

执行结果是:

进队顺序:1 2 3 4 5 6 7 8 9 10
出队顺序:10 9 8 7 6 5 4 3 2 1

可见,出队是谁优先级高,谁先出队。由于输入队列是升序排列,与输出序列刚好相反。

那么现在如果希望按原序列输出,仍然用 priority_queue,该怎么办呢? 先看看 priority_queue 的模板参数定义。

```
template<class T,
    class Cont=vector<T>,
    class Pred=less<Cont::value_type>>
class priority_queue {
    //中间略去
    };
```

因此,示例中 priority_queue<int>pr(a,a+9)相当于 priority_queue<int,vector,less<int>>pr(a,a+9)。所以若想保持按原升序序列输出,只需把 priority_queue<int>pr(a,a+9)改为 priority_queue<int,vector,greater<int>>pr(a,a+9),其他不用修改就可以了。

7.6.3 综合操作示例

【例 7.21】 显示学生信息(学号、姓名、语文、数学),条件是:总成绩由高到低,当总成绩相同时,语文成绩高者优先。

```
//文件名:e7_21.cpp
#include<iostream>
#include<queue>
#include<string>
using namespace std;
class Student
{
    int NO;                                          //学号
    string name;                                     //姓名
    int chinese;                                     //语文
    int math;                                        //数学
public:
    Student(int NO,string name, int chinese, int math)
    {
        this->NO=NO,this->name=name,this->chinese=chinese,this->math=math;
    }
    int GetNO()const {return NO;}
    string GetName()const {return name;}
    int GetChinese() const {return chinese;}
    int GetMath() const {return math;}

    bool operator< (const Student& s)const
```

```cpp
        int sum1=chinese+math;
        int chinese2=s.GetChinese();
        int math2=s.GetMath();
        int sum2=chinese2+math2;

        if(sum1<sum2) return true;
        if((sum1==sum2) && (chinese<chinese2)) return true;
        return false;
    }
};
int main()                                              //测试函数
{
    Student s[]={Student(1001,"zhang",70,80), Student(1002,"li",80,70),
                 Student(1003,"wang",90,85), Student(1004,"zhao",85,75)};
    priority_queue<Student>pr(s, s+4);                  //学生进入优先队列
    cout<<"成绩由高到低(当相同时,语文高优先):"<<endl;
    cout<<"学号\t姓名\t语文\t数学"<<endl;
    while(!pr.empty())
    {
        const Student& t=pr.top();                      //当前成绩最高同学对象
        cout<<t.GetNO()<<"\t"<<t.GetName()<<"\t"<<t.GetChinese()<<"\t"<<t.GetMath()<<endl;
        pr.pop();
    }
    return 0;
}
```

当调用 priority_queue＜Student＞pr(s，s＋4)，其实相当于调用 priority_queue ＜Student,vector＜Student＞，less＜Student＞＞，因此一定要重载基础类 Student 中 operator＜运算符,在该函数内部定义比较算法。

【例 7.22】 设计一个固定大小的优先队列类,即优先队列元素个数只能小于某个值。

分析：标准模板库中 priority_queue 中元素个数没有限制,与题意不符,那么如何既能用上 priority_queue 固有功能,又能加上特有的限制呢？其实在这句话中已经体现出了具体思路：(1) 从 priority_queue 派生出自己定义的优先队列类 FixedPriority。这样,FixedPriority 类就继承了 priority_queue 的固有特性。(2) 在 FixedPriority 类中加入特有的功能,题意中要求限制大小,那么一方面要定义一个成员变量 nLimit,通过构造函数传入优先队列大小,并赋给 nLimit；另一方面要重载 push 函数,如果当前堆栈元素个数小于 nLimit,则把新元素压入堆栈。

    //文件名：e7_22.cpp

```cpp
#include <iostream>
#include <queue>
using namespace std;
template<class T,
    class Cont=vector<T>,
    class Pred=less<typename Cont::value_type>>
class FixedPriority:public priority_queue<T,Cont,Pred>
{
    int nLimit;
public:
    FixedPriority(int nLimit)                //通过构造函数设置优先队列大小
    {
        this->nLimit=nLimit;
    }
    void SetLimitSize(int nLimit)            //或通过函数设置优先队列大小
    {
        this->nLimit=nLimit;
    }
    bool Push(T& t)
    {
        if(nLimit>priority-queue<T,Cont,Pred>::size())
        {
            priority-queue<T,Cont,Pred>::push(t);
            return true;
        }
        return false;
    }
};
int main()                                   //测试
{
    FixedPriority<int>fp(10);                //大小为 10 个元素的优先队列
    for(int i=0; i<15; i++)                  //准备压入 15 个元素
    {
        if(!fp.Push(i))
        {
            cout<<"优先队列已满,第"<<i<<"个元素没有插入"<<endl;  //后 5 个元素没有压入
        }
    }
    return 0;
}
```

同学们可能感觉代码前面的模板参数声明很复杂,不容易记忆,刚开始学的时候,只需

要打开 MSDN 帮助，找到 priority_queue 的模板参数声明，复制过来就可以了。一般来说，在实际中固定化优先队列大小情况非常普遍，希望同学们在本示例基础上加以扩充。

7.7 bitset 容器

因为 C 是一种"接近硬件"的语言，但 C 语言并没有固定的二进制表示法。bitset 可以看做是二进制位的容器，并提供了位的相关操作函数。

7.7.1 常用函数

（1）构造、赋值函数。
- bitset()
- bitset(const bitset&)：复制构造函数。
- bitset(unsigned long val)：由无符号长整型数构建位容器。
- bitset(const string& str, size_t pos=0, size_t n=-1)：由字符串创建位容器。
- bitset& operator=(const bitset&)：赋值操作。

（2）逻辑运算操作（与、或、非）。
- bitset& operator&=(const bitset&)：返回两个位容器"与"后的引用，并修改第一个位容器值。
- bitset& operator|=(const bitset&)：返回两个位容器"或"后的引用，并修改第一个位容器值。
- bitset& operator^=(const bitset&)：返回两个位容器"异或"后的引用，并修改第一个位容器值。
- bitset& operator<<=(size_t)：返回位容器左移 size_t 位后的引用，并修改位容器值。
- bitset& operator>>=(size_t)：返回位容器右移 size_t 位后的引用，并修改位容器值。
- bitset operator<<(size_t n) const：返回位容器左移 size_t 位后的备份。
- bitset operator>>(size_t n) const：返回位容器右移 size_t 位后的备份。
- bitset operator&(const bitset&, const bitset&)：返回两个位容器"与"后的备份。
- bitset operator|(const bitset&, const bitset&)：返回两个位容器"或"后的备份。
- bitset operator^(const bitset&, const bitset&)：返回两个位容器"异或"后的备份。

（3）其他操作函数。
- string to_String()：位容器内容转化成字符串，方便显示。
- size_t size() const：返回位容器大小。
- size_t count() const：返回设置 1 位个数。

第 7 章 通用容器

- bool any() const：是否有位设置 1。
- bool none() const：是否没有位设置 1。
- bool test(size_t n) const：测试某位是否为 1。
- bool operator[](size_t n) const：随机访问位元素。
- unsigned long to_ulong() const：若没有溢出异常,返回无符号长整型数。
- bitset& set()：位容器所有位置 1。
- bitset& flip()：位容器所有位翻转。
- bitset& reset()：位容器所有位置 0。
- bitset& set(size_t n, int val=1)：设置某位为 1 或 0,默认为 1。
- bitset& reset(size_t n)：复位某位为 0。
- bitset flip(size_t n)：翻转某位。

7.7.2 基本操作示例

【例 7.23】 定义位变量简单示例。

```cpp
//文件名：e7_23.cpp
#include<iostream>
#include<string>
#include<bitset>
using namespace std;
int main()
{
    bitset<5>s1;                            //会初始化一个至少具有 5 位的内存空间
    cout<<"初始内存空间内容:"<<s1.to_string()<<endl;
    cout<<"位容器空间(size):"<<s1.size()<<"\t置 1 个数(count):"<<s1.count()<<endl;

    s1.set(2, true);                        //将第 2 位置 1
    cout<<"第 2 位置 1 后[set(2,true)]";
    cout<<"内存空间内容:"<<s1.to_string()<<endl;

    s1[3]=1;                                //另一种设置位方式,相当于数组表示
    cout<<"第 3 位置 1 后[s1[3]=1]";
    cout<<"内存空间内容:"<<s1.to_string()<<endl;

    s1.set();
    cout<<"所有位置 1 后[s1.set()]";
    cout<<"内存空间内容:"<<s1.to_string()<<endl;

    bitset<16>s2(65535);
    cout<<"通过长整型数建立位容器:" <<s2.to_string()<<endl;
```

```
        bitset<5>s3(string("1111101"), 2, 5);
        cout<<"通过字符串建立位容器:"<<s3.to_string()<<endl;
        return 0;
}
```

执行结果是:

初始内存空间内容:00000
位容器空间(size):5   置1个数(count):
第2位置1后[set(2,true)]内存空间内容:00100
第3位置1后[s1[3]=1]内存空间内容:01100
所有位置1后[s1.set()]内存空间内容:11111
通过长整型数建立位容器:1111111111111111
通过字符串建立位容器:11101

(1) 结合结果体会 size 和 count 函数的区别, set 无参、有参函数的区别, 以及可以用[ ] 随机访问位容器的某一位。同时思考一下 reset 无参、有参函数的区别, flip 无参、有参函数的区别。

(2) 当用长整型数建立位容器时, 范围应在$[0, 2^{32})$内, 若位容器设置位数 N 不足以容纳长整型数, 则仅截取假想已是二进制数的低 N 位。例如, bit<2> s4(13), 由于 13 的二进制表示是 1101, 而容器仅两位, 从低位开始填充, 位容器内容为 01。

(3) 用字符串建立位容器 bitset(const string& str, size_t pos=0, size_t n=-1)的含义是: 从字符串第 pos 位置字符开始, 截取 n 个字符填充位容器, 默认设置是填充全字符串。但要注意字符串只能包含"1"或"0"字符, 不能有其他字符。如"10101"正确, 而"12345"错误。

【例 7.24】 位操作函数简单示例。

```
//文件名:e7_24.cpp
#include<iostream>
#include<string>
#includef<bitset>
using namespace std;
int main()
{
        bitset<5>s1(string("01011"));
        bitset<5>s2(string("10010"));
        cout<<"s1:"<<s1.to_string()<<"\t";
        cout<<"s2:"<<s2.to_string()<<endl;
        s1 &= s2;                                        //结果修改了 s1
        cout<<" s1&=s2: "<<s1.to_string()<<endl;
        bitset<5>s3(string("01011"));
        bitset<5>s4(string("10010"));
        bitset<5>s5=s3 & s4;                             //"与"后结果赋值给 s5,s3 及 s4 不变
```

```
        cout<<"s3:"<<s3.to_string()<<"\t";
        cout<<"s4:"<<s4.to_string()<<endl;
        cout<<" s5=s3&s4: "<<s5.to_string()<<endl;
        cout<<" 原 s3: "<<s3.to_string()<<endl;
        return 0;
    }
```

执行结果是：

```
s1: 01011    s2: 10010
         s1&=s2: 00010
s3: 01011    s4: 10010
         s5=s3&s4: 00010
         原 s3: 01011
```

(1) 本示例主要区分 &=、& 操作符的差别。&= 操作符完成"与"操作的同时，修改了第 1 个操作数。& 操作符完成"与"操作，并不修改原操作数。其实从原型定义上也可以看出来：bitset& operator&=(const bitset&)返回当前位容器对象引用，因此当前位容器对象内容可能已发生了变化；bitset operator&(const bitset&，const bitset&)返回位容器对象的备份，且传入的两个与操作位容器对象类型是 const，因此原容器内容一定不发生变化。依此类推，可知 |, |=, ^, ^=, <<, <<=, >>, >>= 的功能差别。

(2) 若两个位容器对象进行各种操作，必须保证位容器大小相同，示例中容器大小均是 5，所以可以进行操作。但如果 bitset<5>s1，bitset<6>s2，则 s1, s2 不能进行各种操作。

【例 7.25】 一个 8 位二进制数，要求高 4 位不变，低 4 位取反。

```
//文件名：e7_25.cpp
#include<iostream>
#include<string>
#include<bitset>
using namespace std;
int main()
{
    bitset<8>b(string("11011101"));
    cout<<"原位容器 b:"<<b.to_string()<<endl;
    for(int i=0; i<4; i++)                              //低 4 位翻转
    {
        b.flip(i);
    }
    cout<<"低 4 位翻转后，b:"<<b.to_string()<<endl;
    return 0;
}
```

主要用到了 flip 函数。还有一个思路：原位容器与另一位容器（固定内容"00001111"）进行异或运算，也是可行的，代码如下：

```
    bitset<8>b("11011101");
    bitset<8>b2("00001111");
    b^=b2;
```

7.7.3 综合操作示例

【例7.26】 编制功能类统计学生每月出勤天数。

分析：若每月都按31天计算，学生有两种状态：要么出勤，要么缺席。故采用bitset位向量来描述是最节省空间的，每月31天状态只用31位，不到4个字节就描述清楚了，代码如下。

```
//文件名：e7_26.cpp
#include<iostream>
#include<string>
#include<bitset>
#include<vector>
using namespace std;

template<size_t N>
class MyAttend
{
    int month;                                          //月份
    bitset<N>b;                                         //出勤位容器
public:
    MyAttend(int month, string strAttend):
        month(month),b(strAttend)
    {
    }
    int GetMonth() {return month;}
    int GetAttendDays() {return b.count();}
};

class Student
{
    string name;                                        //某同学
    vector<MyAttend<31>>v;                              //出勤集合
public:
    Student(string name)
    {
        this->name=name;
    }
    void Add(MyAttend<31>& m)                           //添加某月出勤信息
```

```
            v.push_back(m);
        }
        void ShowAttendDays()                          //显示学生每月出勤情况
        {
            cout<<"姓名:"<<name<<endl;
            cout<<"月份\t出勤天数"<<endl;
            for(int i=0; i<v.size(); i++)
            {
                MyAttend<31>& m=v.at(i);
                int month=m.GetMonth();
                int days=m.GetAttendDays();
                cout<<month<<"\t"<<days<<endl;
            }
        }
    };

    int main()
    {
        Student stud1("zhang");                        //定义 zhang 同学
        string s1="1111100111100111110011111100111";   //1 月份出勤
        string s2="111110011111001111001111100";      //2 月份出勤
        MyAttend<31>m1(1, s1);
        MyAttend<31>m2(2, s2);
        stud1.Add(m1);                                 //添加 zhang 同学 1 月份出勤信息
        stud1.Add(m2);                                 //添加 zhang 同学 2 月份出勤信息
        stud1.ShowAttendDays();                        //显示 zhang 同学出勤信息
        return 0;
    }
```

(1) MyAttend 类成员变量定义了考勤月份变量 month 及一个位容器,用位表示同学每天出勤情况,1 表示出勤,0 表示缺席。在统计出勤天数函数 GetAttendDays 中,用位函数 count 统计 1 的个数就可以了,非常简洁。学生出勤描述类 Student 中,一个学生对应多个月份出勤统计,因此定义了姓名成员变量及出勤的向量 v,它是多个月份学生出勤信息的集合。

(2) 同学们可在此例的基础上加以完善,体会 bitset 容器类的巧妙用处。

【例 7.27】 已知 n 个整型数组,长度都是 10,元素都在[1,20]之间,且均递增排列,无重复数据。试利用 bitset 压缩这些数组,并存入文件中。

分析:假设一个整型数组{1,2,3,4,5,6,7,8,9,10},若按正常方式存入文件,共有 10×4=40 字节。根据要求特点可用一个 20 位的 bitset 容器保存某个数组。若数组中某位值是 5,则把位容器第 5 位置 1,依此类推。20 位相当于 2.5 字节,与原先 40 字节相比,是原先的 1/16。但是文件操作中最小单位是字节，无法读写 2.5 字节，因此位容器选择 24 位大小，这样读写操作就正好是 3 字节了。

```cpp
//文件名：e7_27.cpp
#include<iostream>
#include<fstream>
#include<bitset>
using namespace std;

template<size_t N>
class MyNum
{
public:
    bitset<N>b;
public:
    void Set(int ary[], int nSize)            //把整型数组压缩成位容器
    {
        b.reset();                             //复位位容器
        for(int i=0; i<nSize; i++)
        {
            b.set(ary[i]-1, 1);
        }
    }
};

int main()
{
    int a[6][10]={{1,2,3,4,5,6,7,8,9,10},{1,3,4,5,6,7,8,9,10,12},
                  {2,4,6,8,10,13,15,18,19,20},{1,5,6,8,9,10,12,14,18,20},
                  {3,6,7,10,14,15,16,17,18,19},{1,6,8,9,10,12,15,17,18,19}};
    ofstream out("d:\\data.txt");
    MyNum<24>m;
    for(int i=0; i<6; i++)
    {
        m.Set(a[i], 10);
        out.write((char*)&m.b, 3);
        cout<<i<<"\t";
    }
    out.close();
    return 0;
}
```

(1) MyNum 中定义了位容器变量 b，成员函数 Set 负责把某整型数组转化为位容

器表示。

(2) 在测试 main 中,a[6][10]是待压缩整型数组。主要注意这一行的理解：out.write((char *)&m.b,3)。由于位容器大小是 24,是 3 个字节,因此 write 函数第 2 个参数是 3,表明要写入文件 3 个字节。m.b 表明是位容器变量,位容器变量在内存中是一块连续的区域,和其他基本数据类型相仿,因此可用(char *)&m.b 这种形式。在某些情况下,(int *)&m.b、(float *)&m.b 都是可能的。

那么如何从文件读取,并复原整型变量呢？如下所示。

```
ifstream in("d:\\data.txt");
bitset<24>b;
while(!in.eof())
{
    in.read((char*)&b,3);
    for(int i=0; i<24; i++)
    {
        if(b.test(i))
        {
            cout<<i+1<<"\t";
        }
    }
    cout<<endl;
}
in.close();
```

虽然没有把功能写入类中,但只要把这一段理解好,就可方便地移植到类中。

7.8 集合

set、multiset 都是集合类,差别在于 set 中不允许有重复元素,multiset 中允许有重复元素。

7.8.1 常用函数

(1) 构造函数。
- set(const Pred& comp=Pred(), const A& al=A())：创建空集合。
- set(const set& x)：复制构造函数。
- set(const value_type * first, const value_type * last, const Pred& comp=Pred(), const A& al=A())：复制[first,last]之间元素构成新集合。
- multiset(const Pred& comp=Pred(), const A& al=A())：创建空集合。

- multiset(const multiset& x)：复制构造函数。
- multiset(const value_type * first, const value_type * last, const Pred& comp = Pred(), const A& al = A())：复制[first, last]之间元素构成新集合。

(2) 大小、判断空函数。
- int size() const：返回容器元素个数。
- bool empty() const：判断容器是否空，若返回 true，表明容器已空。

(3) 增加、删除函数。
- pair<iterator, bool> insert(const value_type& x)：插入元素 x。
- iterator insert(iterator it, const value_type& x)：在迭代指针 it 处插入元素 x。
- void insert(const value_type * first, const value_type * last)：插入[first, last)间元素。
- iterator erase(iterator it)：删除迭代指针 it 处元素。
- iterator erase(iterator first, iterator last)：删除[first, last)迭代指针间元素。
- size_type erase(const Key& key)：删除元素值等于 key 的元素。

(4) 遍历函数。
- iterator begin()：返回首元素的迭代器指针。
- iterator end()：返回尾元素后的迭代器指针，而不是尾元素的迭代器指针。
- reverse_iterator rbegin()：返回尾元素的逆向迭代器指针，用于逆向遍历容器。
- reverse_iterator rend()：返回首元素前的逆向迭代器指针，用于逆向遍历容器。

(5) 操作函数。
- const_iterator lower_bound(const Key& key)：返回容器元素大于等于 key 的迭代指针，否则返回 end()。
- const_iterator upper_bound(const Key& key)：返回容器元素大于 key 的迭代指针，否则返回 end()。
- int count(const Key& key) const：返回容器中元素值等于 key 的元素个数。
- pair<const_iterator, const_iterator> equal_range(const Key& key) const：返回容器中元素值等于 key 的迭代指针[first, last)。
- const_iterator find(const Key& key) const：查找功能，返回元素值等于 key 的迭代器指针。
- void swap(set& s)：交换单集合元素。
- void swap(multiset& s)：交换多集合元素。

7.8.2 基本操作示例

【例 7.28】 三种形成集合的方法。

```
//文件名：e7_28.cpp
#include<iostream>
```

```cpp
#include<set>
using namespace std;

void display(multiset<int>& s)
{
    multiset<int>::iterator te=s.begin();
    while(te!=s.end())
    {
        cout<< * te<<"\t";
        te++;
    }
    cout<<endl;
}

int main()
{
    int a[]={5,3,9,3,7,2,9,3};
    multiset<int>myset;
    for(int i=0; i<sizeof(a)/sizeof(int); i++)
    {
        myset.insert(a[i]);
    }

    cout<<"通过函数 insert(T t)添加集合:"<<endl;
    display(myset);                                 //2 3 3 3 5 7 9 9

    cout<<"通过复制构造函数 set(const set& x)创建集合:"<<endl;
    multiset<int>myset2(myset);
    display(myset2);                                //2 3 3 3 5 7 9 9

    cout<<"通过构造函数 set(const value_type * first, const value_type * last)创建集合:"<<endl;
    multiset<int>myset3(a, a+sizeof(a)/sizeof(int));
    display(myset3);                                //2 3 3 3 5 7 9 9
    return 0;
}
```

(1) 可以看出 multiset 集合默认是按升序排列的,这是由于源码中 multiset 模板参数中 Pred=less<Key>决定了键值是按升序排列的。

　　template<class Key, class Pred=less<Key>, class A=allocator<Key>>

(2) multiset 允许有重复元素,如果把 multiset 全部换成 set,则结果为"2 3 5 7 9",没有重复元素,且也按升序排列。

【例 7.29】 equal_range,pair,count,size 用法举例。

//文件名:e7_29.cpp

```cpp
#include<iostream>
#include<set>
using namespace std;
int main()
{
    int a[]={5,3,9,3,7,2,9,3};

    set<int>myset(a, a+sizeof(a)/sizeof(int));              //2 3 5 7 9
    multiset<int>mymultiset(a, a+sizeof(a)/sizeof(int));    //2 3 3 3 5 7 9 9

    pair<set<int>::iterator, set<int>::iterator>rangeset;
    pair<multiset<int>::iterator, multiset<int>::iterator>rangemultiset;

    rangeset=myset.equal_range(3);       //返回集合中等于3的迭代器指针,用pair接收
    rangemultiset=mymultiset.equal_range(3);
                                         //返回多集合中等于3的迭代器指针,用pair接收

    int nCount=myset.count(3);                      //求集合中等于3的元素个数
    int nMultiCount=mymultiset.count(3);            //求多集合中等于3的元素个数

    set<int>::iterator te;                          //集合搜索结果遍历
    vcout<<"set(搜索值等于3的元素): ";
    for(te=rangeset.first; te!=rangeset.second; te++)
    {
        cout<< * te<<"\t";
    }
    cout<<endl;
    cout<<"\t 个数是: "<<nCount<<endl;
    cout<<"\t 总共元素个数是: "<<myset.size()<<endl;

    multiset<int>::iterator it;                     //多集合搜索结果遍历
    cout<<"multiset(搜索值等于3的元素): ";
    for(it=rangemultiset.first; it!=rangemultiset.second; it++)
    {
        cout<< * it<<"\t";
    }
    cout<<endl;
    cout<<"\t 个数是: "<<nMultiCount<<endl;
    cout<<"\t 总共元素个数是: "<<mymultiset.size()<<endl;
    return 0;
}
```

(1) size函数返回容器中总的元素个数,是无参函数; count是有参函数,代表键值,返

回值代表等于键值的元素个数。

(2) equal_range 返回等于键值(首次等于和最后一次等于)的一对迭代指针,封装在系统模板类 pair 对象中,定义如下:

```
template<class T, class U>
    struct pair {
    typedef T first_type;
    typedef U second_type
    T first;
    U second;
    pair();
    pair(const T& x, const U& y);
    template<class V, class W>
        pair(const pair<V, W>& pr);
    };
```

可知 pair 类中两个成员变量为 first,second,类型分别对应模板参数 T,U。由于示例中是对 set 及 multiset 操作的,因此必须定义两个 pair 变量:pair<set<int>::iterator, set<int>::iterator> rangeset 及 pair<multiset<int>::iterator, multiset<int>::iterator> rangemultiset。另外,本例中得到的 pair 对象的 first 成员变量值相当于集合类中的 lower_bound 函数值,second 成员变量值相当于集合类中的 upper_bound 函数值。

7.8.3 综合操作示例

【例 7.30】 编一个集合类,包括并、交、差三种主要功能,不允许有重复数据。并用学生类 Student 加以测试。

分析:很明显要用到 set 类,但 set 本身并没有两集合的并、交、差函数。因此从 set 类派生,增加并、交、差三个主要函数即可。集合类 MySet 及测试用到的 Student 类如下所示。

```
//文件名:e7_30.cpp
#include<iostream>
#include<string>
#include<set>
using namespace std;

class Student
{
private:
    string m_strNO;                              //学号
    string m_strName;                            //姓名
public:
    Student(string no, string name):m_strNO(no),m_strName(name)
```

```cpp
        bool operator<(const Student& s) const        //必须重载,按学号升序排列
        {
            int mark=m_strNO.compare(s.m_strNO);
            return mark<0?true:false;
        }

        string GetNO()const {return m_strNO;}
        string GetName()const {return m_strName;}
};

ostream& operator<<(ostream& os,const Student& s)     //为 Student 对象标准输出做准备
{
    os<<s.GetNO()<<"\t"<<s.GetName();
    return os;
}

template<class T, class Pred=less<T>, class A=allocator<T>>
class MySet:public set<T, Pred, A>
{
public:
    MySet(const T * first, const T * last):set<T>(first, last)
    {
    }
    void add(MySet& second)                           //两集合并
    {
        typename set<T>::iterator te;
        iterator te=second.begin();
        while(te!=second.end())
        {
            set<T>:insert(*te);
            te++;
        }
    }

    void Intersection(MySet& second)                  //两集合交
    {
        set<T>mid;
        typename set<T>::iterator te
        te=set<T>::begin();
        while(te!=set<T>::end())
        {
            if(second.find(*te)!=second.end())
```

```
            }
                mid.insert(*te);
            }
            te++;
        }
        set<T>::swap(mid);
    }

    void Difference(MySet& second)              //两集合差
    {
        set<T>mid;
        typename set<T>::iterator te;
        te=set<T>::begin();
        while(te!=set<T>::end())
        {
            if(second.find(*te)==second.end())
            {
                mid.insert(*te);
            }
            te++;
        }
        set<T>::swap(mid);
    }


    void Show()                                 //集合显示函数
    {
        typename set<T>::iterator te;
        te=set<T>::begin();
        while(te!=set<T>::end())
        {
            cout<<*te<<endl;
            te++;
        }
    }
};

int main()
{
    Student s[2]={{Student("1001","zhangsan")},
                  {Student("1002","lisi")}};
    Student t[3]={{Student("1001","zhangsan")},
                  {Student("1003","wangwu")},
                  {Student("1004","zhaoliu")}};
```

```
            MySet<Student>m1(s, s+2);
            MySet<Student>m2(t, t+3);

            cout<<"原始 m1 集合:"<<endl;
            m1.Show();
            cout<<"原始 m2 集合:"<<endl;
            m2.Show();

            cout<<"m1 并 m2 为:"<<endl;
            m1.add(m2);       //求 m1 并 m2,若此位置改为 Intersection 可求交,改为 Difference 可求差
            m1.Show();
            return 0;
    }
```

(1) 在 MySet 类中: add 函数完成了集合的并运算,主要是通过系统提供的 insert 函数来完成的; Intersection(MySet& second)函数完成了集合的交运算,先定义了一个空的集合临时变量 set<T> mid,之后遍历当前集合中的各元素,若当前遍历元素在第二个集合中存在,则作为两个集合的共有元素加入临时集合变量 mid 中,最后利用系统提供的 swap 函数交换临时集合变量 mid 及当前集合对象; Difference(MySet& second)函数完成了集合的差运算,算法与上述的交运算相似,就不再论述了。

(2) 必须重载 Student 类中的 operator<运算符,保证集合按照学号升序排列,也就是说除了基本数据类型(int,float 等)外,集合中元素是类对象的,必须重载类中的 operator<运算符。

7.9 映射

常用的映射类是 map、multimap。在前述的各个容器中,仅保存着一样东西,但是在映射中将会得到两样东西:关键字以及作为对关键字进行查询得到的结果值,即一对值<Key,Value>。map 单映射中,Key 与 Value 是一对一的关系;multimap 多映射中,Key 与 Value 可以是一对多的关系。

7.9.1 常用函数

(1) 构造函数。
- map(const Pred& comp=Pred(), const A& al=A()):创建空映射。
- map(const map& x):复制构造函数。
- map(const value_type * first, const value_type * last, const Pred& comp=Pred(), const A& al=A()):复制[first,last]之间元素构成新映射。
- multimap(const Pred& comp=Pred(), const A& al=A()):创建空映射。

- multimap(const multimap& x)：复制构造函数。
- multimap(const value_type * first, const value_type * last, const Pred& comp = Pred(), const A& al = A())：复制[first,last)之间元素构成新映射。

(2) 大小、判断空函数。
- int size() const：返回容器元素个数。
- bool empty() const：判断容器是否空，若返回 true，表明容器已空。

(3) 增加、删除函数。
- iterator insert(const value_type& x)：插入元素 x。
- iterator insert(iterator it, const value_type& x)：在迭代指针 it 处插入元素 x。
- void insert(const value_type * first, const value_type * last)：插入[first,last)间元素。
- iterator erase(iterator it)：删除迭代指针 it 处元素。
- iterator erase(iterator first, iterator last)：删除[first,last)迭代指针间元素。
- size_type erase(const Key& key)：删除键值等于 key 的元素。

(4) 遍历函数。
- iterator begin()：返回首元素的迭代器指针。
- iterator end()：返回尾元素后的迭代器指针，而不是尾元素的迭代器指针。
- reverse_iterator rbegin()：返回尾元素的逆向迭代器指针，用于逆向遍历容器。
- reverse_iterator rend()：返回首元素前的逆向迭代器指针，用于逆向遍历容器。

(5) 操作函数。
- const_iterator lower_bound(const Key& key)：返回键值大于等于 key 的迭代指针，否则返回 end()。
- const_iterator upper_bound(const Key& key)：返回键值大于 key 的迭代指针，否则返回 end()。
- int count(const Key& key) const：返回容器中键值等于 key 的元素个数。
- pair<const_iterator, const_iterator> equal_range(const Key& key) const：返回容器中键值的迭代指针[first, last)。
- const_iterator find(const Key& key) const：查找功能，返回键值等于 key 的迭代器指针。
- void swap(map& s)：交换单映射元素。
- void swap(multimap& s)：交换多映射元素。

(6) 特殊函数。
reference operator[](const Key& k)：仅用在单映射 map 类中，可以以数组的形式给映射添加键-值对，并可返回值的引用。

7.9.2 基本操作示例

【例 7.31】 两种常用的形成映射方法。

//文件名：e7_31.cpp

```cpp
#include<iostream>
#include<string>
#include<map>
using namespace std;
void Display(map<int, string>& m)
{
    map<int, string>::iterator te=m.begin();
    while(te!=m.end())
    {
        cout<<(*te).first <<"\t"<<(*te).second<<endl;
        te++;
    }
}
int main()
{
    map<int, string>mymap;
    pair<int, string>s1(1, "zhangsan");
    pair<int, string>s2(3, "lisi");
    pair<int, string>s3(6, "wangwu");
    pair<int, string>s4(5, "zhaoliu");
    pair<int, string>s5(1, "zhangsan");
    mymap.insert(s1);
    mymap.insert(s2);
    mymap.insert(s3);
    mymap.insert(s4);
    mymap.insert(s5);
    cout<<"通过 insert 函数创建:"<<endl;
    Display(mymap);
    cout<<"通过复制构造函数创建:"<<endl;
    map<int, string>mymap2(mymap);
    Display(mymap2);
    return 0;
}
```

（1）可以看出 map 映射默认是按键值升序排列的，这是由于源码中 map 模板参数中，Pred＝less＜Key＞决定了键值是按升序排列的。

  template<class Key, class T, class Pred=less<Key>, class A=allocator<T>>

（2）map 不允许有重复键值，本例中有 5 个 pair 对象，且 s1、s5 有相同的键值，则在 map 容器中只保留先存入的 s1 对象，s5 则不能保存。如果把 map 换成 multimap，则可以保存 s5 了。

（3）前面讲过 pair 是一个两个模板参数的模板类，正好符合映射类中的"键-值"映射要求,因此映射类要求保存的都是 pair 对象。在本示例 Display 显示函数中,获得的迭代指针都是 pair<int,string> * 的,所以(*te).first 及(*te).second 中的 first、second 变量均是 pair 类中的成员变量,依据映射类规定: * first 表示键值, * second 表示映射的值。

【例 7.32】 单映射 map 中 operator[]用法。

```
//文件名: e7_32.cpp
#include<iostream>
#include<string>
#include<map>
using namespace std;
int main()
{
    map<string, string>mymap;
    mymap["1-1"]="元旦";                //通过赋值形式添加映射,键在[]内,值在等号右侧
    mymap["5-1"]="五一国际劳动节";
    mymap["7-1"]="党的生日";
    mymap["8-1"]="建军节";
    mymap["10-1"]="国庆节";
    string s=mymap["1-1"];              //通过 operator[]完成查询功能,赋给 string 变量 s
    if(s.size()>0)
    {
        cout<<"1-1是:"<<s<<endl;
    }
    else
    {
        cout<<"1-1没有登记"<<endl;
    }
    s=mymap["6-1"];
    if(s.size()>0)
    {
        cout<<"6-1是:"<<s<<endl;
    }
    else
    {
        cout<<"6-1没有登记"<<endl;
    }
    return 0;
}
```

(1) 例如"mymap["1-1"]="元旦""是"数组"形式,但不用指明大小,也不用 new 分配空间。[]内的是键值,由于本例是 map<string,string>,所以键是 string 类型的。

(2) 可以利用行如"string s=mymap["1-1"]"完成查询。对本例而言,如果有查询结

果,则字符串 s 长度不为 0,反之为 0。

7.9.3 综合操作示例

【例 7.33】 假设公司雇员属性有雇员姓名(没有重复的姓名)、部门名称。编制管理雇员的集合类,仅包含:(1)添加雇员功能;(2)显示功能,要求按部门名称升序排列,若部门名相同,则按姓名升序排列。

分析:应该利用集合类,在添加雇员时,直接完成先按部门升序排列,再按姓名升序排列,即在恰当的位置重载 operator< 运算符,完成自定义排序规则功能。由于一个部门可以有许多雇员,因此应当采用 multiset 类。设雇员基本类为 CEmployee,雇员集合类为 CManage,则代码如下所示。

```
//文件名:e7_33.cpp
#include<iostream>
#include<string>
#include<set>
using namespace std;
class CEmployee                                          //雇员基础类
{
private:
    string name;
    string departname;
public:
    CEmployee(string name, string departname)
    {
        this->name=name;
        this->departname=departname;
    }
    bool operator< (const CEmployee& e) const            //定义添加接口
    {
        bool mark= (departname.compare(e.departname)<0)?true:false;
                                                         //部门升序
        if(departname.compare(e.departname)==0)
        {
            mark= (name.compare(e.name)<0)?true:false;   //姓名升序
        }
        return mark;
    }
    string GetName()const{return name;}
    string GetDepart()const{return departname;}
};
class CManage                                            //雇员集合维护类
```

```
    multiset<CEmployee>myset;
public:
    bool Add(CEmployee& e)
    {
        myset.insert(e);
        return true;
    }
    void Show()
    {
        multiset<CEmployee>::iterator te=myset.begin();
        while(te!=myset.end())
        {
            const CEmployee& obj= * te;
            cout<<obj.GetDepart()<<"\t"<<obj.GetName()<<endl;
            te++;
        }
    }
};

int main()                              //测试 main 函数
{
    CEmployee e1("zhangsan","人力部");
    CEmployee e2("zhouqi","装配部");
    CEmployee e3("wangwu","制造部");
    CEmployee e4("zhaoliu","制造部");
    CEmployee e5("lisi","装配部");
    CEmployee e6("tianjiu","制造部");

    CManage manage;
    manage.Add(e1);manage.Add(e2);
    manage.Add(e3);manage.Add(e4);
    manage.Add(e5);manage.Add(e6);
    manage.Show();
    return 0;
}
```

【例 7.34】 编一个同义词字典功能类，每个单词后面跟着它的同义词，示例如下：

one unique single
correct true right
near close
……

需要解决的问题是：形成上述的同义词字典，根据给定的单词查出相应的同义词。

分析：一般来说形成的词典文件是按照字母顺序排列的，又由于一个单词可能有很多的同义词，因此 multimap 多映射类更适于解决此类问题。设基础单词类为 CWord，同义词集合管理类为 CWordManage。

```cpp
//文件名：e7_34.cpp
#include<iostream>
#include<sstream>
#include<string>
#include<vector>
#include<map>
using namespace std;

class CWord                                         //同义词基础类
{
private:
    string mainword;                                //一个单词
    vector<string>vecword;                          //可以有多个同义词
public:
    CWord(string strLine)
    {
        istringstream in(strLine);                  //转化成字符串流
        in>>mainword;                               //第一个单词是关键词
        string mid="";
        while(!in.eof())                            //其余词是第一个词的近义词
        {
            in>>mid;
            vecword.push_back(mid);
        }
    }
    string GetMainWord(){return mainword;}
    void Show()
    {
        cout<<endl;
        cout<<"单词是："<<"\t"<<mainword<<endl;
        cout<<"同义词是："<<"\t";
        for(int i=0; i<vecword.size(); i++)
        {
            cout<<vecword[i]<<"\t";
        }
        cout<<endl;
    }
};
```

```cpp
class CWordManage                              //同义词集合类
{
    multimap<string,CWord>mymap;
public:
    bool Add(string strLine)                   //形成集合同义词映射
    {
        CWord word(strLine);
        pair<string,CWord>p(word.GetMainWord(),word);
        mymap.insert(p);
        return true;
    }

    void Show(string strFind)                  //显示待查的同义词
    {
        multimap<string,CWord>::iterator itfind=mymap.find(strFind);
        if(itfind!=mymap.end())                //如果有同义词
        {
            CWord& obj=(*itfind).second;
            obj.Show();
        }
        else                                    //如果没有同义词
        {
            cout<<strFind<<"字典里没有记录同义词"<<endl;
        }
    }

    void Show()                                 //显示字典中所有同义词信息
    {
        multimap<string,CWord>::iterator te=mymap.begin();
        while(te!=mymap.end())
        {
            CWord& obj=(*te).second;
            obj.Show();
            te++;
        }
    }
};

int main()                                      //仿真测试
{
    string s[5]={string("one single unique"),string("correct true right"),
                 string("near close"), string("happy please"),
                 string("strong powerful")};
```

```
        CWordManage manage;
        for(int i=0; i<5; i++)                    //形成同义词词典
        {
            manage.Add(s[i]);
        }

        manage.Show();                            //显示同义词词典所有信息
        cout<<"****************************************************"<<endl;
        manage.Show("near");                      //查询 near 的同义词
        cout<<"****************************************************"<<endl;
        manage.Show("good");                      //查询 near 的同义词
        return 0;
    }
```

(1) 同义词基础类 CWord 定义了两个成员变量：string mainword 和 vector<string> vecword。表明一个单词有多个同义词，vecword 定义成向量 vector 类型，不要定义成静态的数组类型。构造函数 CWord(string strLine)的功能是按空格解析一行字符串，第 1 个字符串作为关键字保存在 mainword 成员变量中，其余拆分出来的字符串保存在向量 vecword 中。

(2) 同义词集合类 CWordManage 是 CWord 对象的集合。Add(string strLine)函数功能是添加一个新的同义词对象 CWord。切记必须把 CWord 对象封装在系统的 pair 对象中，这样才能用 multimap 中的 insert 函数把新的 CWord 对象添加进去。

(3) main 函数中仅是用字符串数组作了简单的仿真，同学们可以通过文件或数据库形成词典内容庞大的 CWordManage 对象。

7.10 再论迭代器

STL 有一些可用的预迭代器，第 3 章介绍了 istream_iterator 及 ostream_iterator，还有一些预迭代器与本章所述各容器相关。

1. 插入迭代器（仅列出了构造函数，当然它们的模板参数都是容器 Cont）

- back_insert_iterator(Cont& x)：容器后插迭代器。
- front_insert_iterator(Cont& x)：容器前插迭代器。
- insert_iterator(Cont& x,Cont::iterator it)：容器某指定位置插入迭代器。

插入迭代器所做的事情就是改变操作符 operator＝的实现来替代赋值操作。构造函数都使用一个基本序列容器对象(vector、list 等)作为其参数。back_insert_iterator 通过调用 push_back 函数完成后插功能，即容器中有 push_back 函数才支持该迭代器；front_insert_iterator 通过调用 push_front 函数完成前插功能，即容器中有 push_front 函数才支持该迭代器;insert_iterator 通过调用 insert 函数完成插入操作,插入位置由 Cont::iterator it 迭代指针决定,它对所有基本序列容器都适用。

另外,还有两个迭代模板函数经常要用到。

- back_insert_iterator<Cont> back_inserter(Cont& x):返回容器 x 的后插迭代器。
- front_insert_iterator<Cont> front_inserter(Cont& x):返回容器 x 的前插迭代器。

这两个函数的功能与 back_insert_iterator、front_insert_iterator 功能是一致的,只不过从编程角度来说可以少写一些代码。但要注意:back_inserter、front_inserter 是函数,而 back_insert_iterator、front_insert_iterator 是类。

【例 7.35】 三个插入迭代器示例。

```
//文件名:e7_35.cpp
#include<iostream>
#include<list>
#include<iterator>
using namespace std;
void display(list<int>& v)
{
    for(list<int>::iterator it=v.begin(); it!=v.end(); it++)
    {
        cout<< * it<<"\t";
    }
    cout<<endl;
}
int main()
{
    list<int>v;
    back_insert_iterator<list<int>> backit(v);
    * backit++=1;
    * backit++=2;
    display(v);                                     //1 2
    * back_inserter(v)=3;
    * back_inserter(v)=4;
    display(v);                                     //1 2 3 4
    front_insert_iterator<list<int>> frontit(v);
    * frontit++=5;
    * frontit++=6;
    display(v);                                     //6 5 1 2 3 4
    * front_inserter(v)++=7;
```

```
        *front_inserter(v)++=8;
        display(v);                                            //8 7 6 5 1 2 3 4

        list<int>::iterator it=v.begin();
        for(int i=0; i<3; i++)
        {
            it++;
        }                                                      //获得 list 容器中第 4 个元素的迭代指针
        insert_iterator<list<int>>insertit(v, it);             //在第 4 个元素前插入 9
        *insertit++=9;                                         //8 7 6 9 5 1 2 3 4
        display(v);
        return 0;
}
```

本例是以 list 容器为示例的。由于有 push_back、push_front 函数，因此支持 back_insert_iterator、front_insert_iterator 迭代器，当然更支持 insert_iterator 迭代器。各迭代器应用方法主要见代码黑体部分。同学们自己分析一下结果原因，特别注意 back_inserter、front_inserter 函数的用法。

2. 逆向迭代器（仅列出了构造函数）

- reverse_iterator(Ranit x)：Ranit 表示随机迭代器。
- reverse_bidirectional_iterator(Bidit x)：Bidit 表示双向迭代器。

逆向迭代器的主要目的是逆向遍历容器元素序列，它一般需要两个模板参数，第一个参数是迭代器类型，第二个参数是容器元素类型，例如创建一个 vector 整型元素逆向迭代器为（其中假设 v 表示已创建的 vector＜int＞容器）reverse_iterator＜vector＜int＞::iterator, int＞ first(v.begin())。

那么，如何应用这两个迭代器呢？一般来说，若容器中迭代器是随机迭代器，则应用 reverse_iterator；若容器中迭代器是双向迭代器，则应用 reverse_bidirectional_iterator。例如，vector 容器中迭代器是随机迭代器，则首选 reverse_iterator；list 容器中迭代器是双向迭代器，则首选 reverse_bidirectional_iterator。

【例 7.36】 两个逆向迭代器示例。

```
//文件名：e7_36.cpp
#include<iostream>
#include<list>
#include<vector>
#include<iterator>
using namespace std;

template<class reverse_iter>
void reverse_display(reverse_iter first, reverse_iter last)    //通用模板逆向显示函数
```

```
{
    while(first!=last)
    {
        cout<< * first<<"\t";
        first++;
    }
    cout<<endl;
}
int main()
{
    vector<int>v;
    list<int>    l;
    for(int i=1; i<=5; i++)
    {
        v.push_back(i);                        //v:1 2 3 4 5
        l.push_back(i+5);                      //l:6 7 8 9 10
    }

    cout<<"vector 元素逆向显示:"<<endl;
    reverse_iterator<vector<int>::iterator, int>first(v.end());
    reverse_iterator<vector<int>::iterator, int>last(v.begin());
    reverse_display(first, last);              //5 4 3 2 1

    cout<<"list 元素逆向显示:"<<endl;
    reverse_bidirectional_iterator<list<int>::iterator, int>first2(l.end());
    reverse_bidirectional_iterator<list<int>::iterator, int>last2(l.begin());
    reverse_display(first2, last2);            //10 9 8 7 6
    return 0;
}
```

程序创建了 vector、list 两个整型容器对象。vector 容器值为 1,2,3,4,5,list 容器值为 6,7,8,9,10。然后,vector 采用 reverse_iterator 逆向迭代器完成了容器的逆向显示"5 4 3 2 1",list 采用 reverse_bidirectional_iterator 完成了容器的逆向显示"10 9 8 7 6"。

可以做一下测试:如果把对 vector 容器采用的 reverse_iterator 改为 reverse_bidirectional_iterator,其他不变,程序结果仍然正确;但如果把 list 容器的 reverse_bidirectional_iterator 改为 reverse_iterator,则程序不能通过编译。同学们可以深入地思考一下其原因。

3. 迭代器函数

有两个常用的迭代器函数,如下所示。

• advance:前移迭代器指针。

```
template<class InIt, class Dist>
```

```
void advance(InIt& it, Dist n);
```

相当于 InIt += n，它是通过 n 步 ++InIt 得到最终迭代器指针的。

- distance：计算迭代器起止指针间有多少元素。

```
template<class InIt, class Dist>
ptrdiff_t distance(InIt first, InIt last);
```

【例 7.37】 已知整型 list 容器序列{1,2,3,4,5,6}。求：(1)元素 3 是第几个元素；(2)显示 3 前面第 2 个元素。

```cpp
//文件名：e7_37.cpp
#include<iostream>
#include<list>
#include<algorithm>
#include<iterator>
using namespace std;
int main()
{
    int a[]={1,2,3,4,5,6};
    list<int>l(a, a+6);
    cout<<"list 容器元素：";
    copy(l.begin(), l.end(), ostream_iterator<int>(cout, "\t"));
    cout<<endl;

    list<int>::iterator mid=find(l.begin(), l.end(), 3);

    cout<<"元素 3 位置："<<distance(l.begin(), mid)<<endl;

    advance(mid, 2);
    cout<<"元素 3 前面第 2 个元素是："<< *mid<<endl;
    return 0;
}
```

思路是先用 find 函数查找元素 3 所在的迭代指针 mid；然后用 distance(l.begin(), mid)就获得了 3 是第几个元素；再用 advance(mid,2)就获得了 3 前面第 2 个元素的迭代器指针，显示其值也就容易了。

第 8 章 非变异算法

非变异算法不直接改变其操作的数据结构的元素。通常,它们查找数据结构中的元素,检查序列元素是否满足某函数式,计算序列元素满足某条件的个数等。表 8.1 所示为按照功能划分的非变异函数。

表 8.1 非变异函数列表

序号	功能	函数名称	说 明
1	循环	for_each	遍历容器元素,对每个元素执行相同的函数操作
2	查询	find	在单迭代器序列中找出某个值第一次出现的位置
		find_if	在单迭代器序列中找出符合某谓词的第一个元素
		find_first_of	在双迭代器序列中找出子序列中某元素第一次出现的位置
		adjacent_find	在单迭代器序列中找出第一次相邻值相等元素的位置
		find_end	在双迭代器序列中找出一子序列最后一次出现的位置
		search	在双迭代器序列中找出一个子序列第一次出现的位置
		search_n	在单迭代器序列中找出一个值连续 n 次出现的位置
3	计数	count	在序列中统计某个值出现的次数
		count_if	在序列中统计与某谓词(表达式)匹配的次数
4	比较	equal	两个序列中的对应元素都相同时为真
		mismatch	找出两个序列相异的第一个元素

8.1 循环

8.1.1 主要函数

for_each():遍历容器元素,对每个元素执行相同的函数操作。
函数原型如下所示。

```
template<class InIt, class Fun>
    Fun for_each(InIt first, InIt last, Fun f);
```

参数说明:
- InIt：可以是数组或输入迭代器类。first 表示数组的起始元素指针或输入迭代器的起始迭代指针，last 表示数组结束元素指针或输入迭代器的终止迭代指针。
- f：可以是全局函数或一元函数。

该模板函数的含义是[first，last)之间的每个元素，作为函数 f 的参数传入并执行。

8.1.2 示例分析

【例 8.1】 打印向量中每个整型元素的立方。

```cpp
//文件名：e8_1.cpp
#include<iostream>
#include<vector>
#include<algorithm>
using namespace std;

void PrintCube(int n)             //打印 n 的立方
{
    cout<<n*n*n<<" ";
}
int main()
{
    const int VECTOR_SIZE=8;
    typedef vector<int> IntVector;              //为向量类起一个别名
    typedef IntVector::iterator IntVectorIt;    //为向量类迭代器起一个别名

    IntVector Numbers(VECTOR_SIZE);    //初始化向量类,大小为 VECTOR_SIZE(8)
    IntVectorIt start, end, it;        //定义向量类迭代指针

    for (int i=0; i<VECTOR_SIZE; i++)  //通过数组方式给向量类赋值
        Numbers[i]=i+1;

    start=Numbers.begin();             //start 表示迭代器起始指针
    end=Numbers.end();                 //end 表示迭代器结束指针

    cout<<"Numbers { ";                //打印向量中的各个元素
    for(it=start; it!=end; it++)
        cout<< *it<<" ";
```

第8章 非变异算法

```
        cout<<" }\n"<<endl;

        for_each(start, end, PrintCube);
                        //[start,end)间每个元素作为 PrintCube 的参数传入并执行
        cout<<"\n";
        return 0;
}
```

该程序功能是先给向量赋初值,调用 for_each 函数,显示向量中各元素的立方值。需要着重理解两点:

(1) for_each 函数各参数的含义。start,end 表示向量的起始迭代指针、结束迭代指针,不是具体的值,比如 for_each(start[0],end,PrintCube)是错误的,因为 start[0]表示的是向量中第一个元素的值 1。但 for_each(&start[0],end,PrintCube)是正确的,因为 &start[0]表示的是第一个元素的地址。

(2) PrintCube 函数必须有且只有一个参数,且参数类型与向量的模板类型一致。

【例 8.2】 求整数向量的和、最大值、最小值。

分析:由于必须遍历整型向量的每个元素,因此可以应用 for_each 函数。当然,可以应用例 8.1 中定义全局函数的方法完成所需功能,但更好的方法是定义一个类,如下所示。

```
//文件名: e8_2.cpp
#include<iostream>
#include<algorithm>
using namespace std;
class PrintInfo
{
private:
    int nSum;
    int nMax;
    int nMin;
    int count;
public:
    PrintInfo():count(0),nSum(0) {}
    int GetSum() {return nSum;}
    int GetMax() {return nMax;}
    int GetMin() {return nMin;}
    void operator() (int x)
    {
        if(count ==0)
        {
            nMax=x;
            nMin=x;
```

```
                }
                else
                {
                    if(nMax < x)
                    {
                        nMax=x;
                    }
                    if(nMin>x)
                    {
                        nMin=x;
                    }
                    nSum+=x;
                    count++;
                }
            };

int main()
{
    int A[]={1, 4, 2, 8, 5, 7};
    const int N=sizeof(A)/sizeof(int);

    PrintInfo P=for_each(A, A+N, PrintInfo());
    cout<<"总和是:"<<P.GetSum()<<endl;
    cout<<"最大值:"<<P.GetMax()<<endl;
    cout<<"最小值:"<<P.GetMin()<<endl;
    return 0;
}
```

对该程序应强化记忆或理解以下几点：

(1) 必须重载 PrintInfo 的 operator()运算符，本例中即 void operator()(int x)，在此函数内完成了求和、最大值、最小值的功能。

(2) 主程序中 for_each(A，A+N，PrintInfo<int>())运行时，程序先执行 PrintInfo<int>()，即调用了 PrintInfo 类中不含参数的构造函数，然后数组中的每一项依次调用 PrintInfo 类中的 void operator()(int x)函数，完成相应功能。

(3) 对主程序中 PrintInfo<int> P=for_each(A，A + N，PrintInfo<int>())等号左边 PrintInfo<int> P 的理解。可从 for_each 原型看出：

```
template<class InIt, class Fun>
    Fun for_each(InIt first, InIt last, Fun f);
```

返回值可以是函数对象，因此本例中等号左侧定义了 PrintInfo& P 引用对象。当然，只有当[first，last)中所有迭代值都执行了 f 函数，才最终返回函数对象。

当然,本例只能求整型向量的和、最大值、最小值,若推广成更普通的类型,毫无疑问要采用模板传参技术,对 for_each 函数而言,最好采用 STL 提供的一元函数模板技术,这样可方便程序 STL 技术的延续性。功能类代码如下。

```
#include<iostream>
#include<algorithm>
#include<functional>
using namespace std;

template <class T, class _outPara>
class PrintInfo:public unary_function<T, _outPara>
{
private:
    T nSum;
    T nMax;
    T nMin;
    int count;
public:
    PrintInfo():count(0),nSum(0) {}
    T GetSum() {return nSum;}
    T GetMax() {return nMax;}
    T GetMin() {return nMin;}

    _outPara operator() (T x)
    {
        if(count ==0)
        {
            nMax=x;
            nMin=x;
        }
        else
        {
            if(nMax <x)
            {
                nMax=x;
            }
            if(nMin >x)
            {
                nMin=x;
            }
        }
        nSum+=x;
```

```
        count++;
    }
};
int main()
{
    float A[]={1.5, 4.2, 2.6, 8.9, 5.7, 7.1};
    const int N=sizeof(A)/sizeof(int);
    PrintInfo<float, void>& P=for_each(A, A+N, PrintInfo<float, void>());
    cout<<"总和是:"<<P.GetSum()<<endl;
    cout<<"最大值:"<<P.GetMax()<<endl;
    cout<<"最小值:"<<P.GetMin()<<endl;
    return 0;
}
```

8.2 查询

8.2.1 主要函数

查询操作是应用最广的操作，STL 主要提供了以下查询函数。
- find()：在单迭代器序列中找出某个值第一次出现的位置。
- find_if()：在单迭代器序列中找出符合某谓词的第一个元素。
- find_first_of()：在双迭代器序列中找出一子序列中某元素第一次出现的位置。
- adjacent_find()：在单迭代器序列中找出第一次相邻值相等元素的位置。
- find_end()：在双迭代器序列中找出一子序列最后一次出现的位置。
- search()：在双迭代器序列中找出一子序列第一次出现的位置。
- search_n()：在单迭代器序列中找出一个值连续 n 次出现的位置。

各个函数原型如下所示。

① find

原型：

```
template<class InIt, class T>
    InIt find(InIt first, InIt last, const T& val);
```

参数说明：
- InIt：输入迭代器，first 表示起始元素的迭代器指针，last 表示结束元素的迭代器指针。
- T：模板类型参数。

该函数是查询[first, last)间迭代器对应的元素值是否有等于 val 的，若有则返回其迭代器指针；若无则返回 last。可知查询元素的个数范围是 N∈[0, last−first)，由于要判定

第8章 非变异算法

*(first+N)==val，因此模板 T 对应的类必须重载运算符"operator=="。

② find_if

原型：

template<class InIt, class Pred>
 InIt **find_if**(InIt first, InIt last, Pred pr);

参数说明：

- InIt：输入迭代器，first 表示起始元素的迭代器指针，last 表示结束元素的迭代器指针。
- Pred：普通全局函数或一元函数对象，返回值是 bool 类型。

该函数是查询[first, last)间迭代器对应的元素 *(first+i)，若 pr(*(first+i))返回 true，则返回此时的迭代器指针，表明满足条件的元素已找到；若没有找到则返回 last。

③ find_first_of

原型：

template<class FwdIt1, class FwdIt2>
 FwdIt1 find_first_of(FwdIt1 first1, FwdIt1 last1,FwdIt2 first2, FwdIt2 last2);
template<class FwdIt1, class FwdIt2, class Pred>
 FwdIt1 **find_first_of**(FwdIt1 first1, FwdIt1 last1,FwdIt2 first2, FwdIt2 last2, Pred pr);

参数说明：

- FwdIt1,FwdIt2：前向迭代器，first 表示起始元素的迭代器指针，last 表示结束元素的迭代器指针。
- Pred：二元全局函数或函数对象。

第一个原型含义是：若第一个前向迭代器[first1，last1)间第 N 个元素与第二个前向迭代器[first2，last2)间某元素相等，且 N 最小，则返回 first1+N。表明第一个前向迭代器 FwdIt1 中有元素与第二个前向迭代器 FwdIt2 中的元素相等，否则返回 last1。

第二个原型与第一个类似，只不过要定义预判定函数 pr(*(first1+N)，*(first2+M))。

④ adjacent_find

原型：

template<class FwdIt>
 FwdIt **adjacent_find**(FwdIt first, FwdIt last);
template<class FwdIt, class Pred>
 FwdIt **adjacent_find**(FwdIt first, FwdIt last, Pred pr);

参数说明：

- FwdIt：前向迭代器，first 表示起始元素的迭代器指针，last 表示结束元素的迭代器指针。
- Pred：二元全局函数或函数对象。

第一个原型含义是：若前向迭代器 FwdIt 中存在第 N 个元素，有 *(first+N)=*(first+N+1)，且 N 最小，则表明有两个相邻元素是相等的，返回(first+N)，否则返回 last。

第二个原型与第一个类似，只不过要定义预判定函数 pr(*(first+N),*(first+N+1))。

⑤ find_end

原型：

```
template<class FwdIt1, class FwdIt2>
    FwdIt1 find_end(FwdIt1 first1, FwdIt1 last1,FwdIt2 first2, FwdIt2 last2);
template<class FwdIt1, class FwdIt2, class Pred>
    FwdIt1 find_end(FwdIt1 first1, FwdIt1 last1,FwdIt2 first2, FwdIt2 last2, Pred pr);
```

参数说明：

- FwdIt1，FwdIt2：前向迭代器，first 表示起始元素的迭代器指针，last 表示结束元素的迭代器指针。
- Pred：二元全局函数或函数对象。

第一个原型含义是：若前向迭代器 FwdIt1 从第 N 个元素开始：*(first1+N)=*(first2+0)，*(first1+N+1)=*(first2+1)，…，*[first1+(last2−first2−1)]=*[first2+(last2−first2−1)]，且 N 最大，则返回(first1+N)，否则返回 last1。即返回在 FwdIt1 元素中最后一次完全与 FwdIt2 序列元素匹配的开始位置。

第二个原型与第一个类似，只不过要定义预判定函数 pr(*(first1+N+M),*(first2+N+M))。

⑥ search

原型：

```
template<class FwdIt1, class FwdIt2>
    FwdIt1 search(FwdIt1 first1, FwdIt1 last1,FwdIt2 first2, FwdIt2 last2);
template<class FwdIt1, class FwdIt2, class Pred >
    FwdIt1 search(FwdIt1 first1, FwdIt1 last1,FwdIt2 first2, FwdIt2 last2,Pred pr);
```

参数说明：

- FwdIt1，FwdIt2：前向迭代器，first 表示起始元素的迭代器指针，last 表示结束元素的迭代器指针。
- Pred：二元全局函数或函数对象。

第一个原型含义是：若前向迭代器 FwdIt1 从第 N 个元素开始：*(first1+N)=*(first2+0)，*(first1+N+1)=*(first2+1)，…，*[first1+(last2−first2−1)]=*[first2+(last2−first2−1)]，且 N 最小，则返回(first1+N)，否则返回 last2。即返回在 FwdIt1 元素中首次完全与 FwdIt2 序列元素匹配的开始位置。

第二个原型与第一个类似，只不过要定义预判定函数 pr(*(first1+N+M),*(first2+M))。

⑦ search_n

原型：

第 8 章 非变异算法

```
template<class FwdIt, class Dist, class T>
    FwdIt search_n(FwdIt first, FwdIt last,Dist n, const T& val);
template<class FwdIt, class Dist, class T, class Pred>
    FwdIt search_n(FwdIt first, FwdIt last,Dist n, const T& val, Pred pr);
```

参数说明：
- FwdIt：前向迭代器，first 表示起始元素的迭代器指针，last 表示结束元素的迭代器指针。
- n：整型变量，表示大小。
- val：待比较的值。
- Pred：二元全局函数或函数对象。

第一个原型含义是：在前向迭代器 FwdIt 中，从第 N 个元素开始连续的 n 个元素满足：*(first＋N)＝val，*(first＋1)＝val，…，*(first＋N＋n)＝val，且 N 最小，则返回 *(first＋N)，否则返回 last。

第二个原型与第一个类似，只不过要定义预判定函数 pr(*(first1＋N＋M)，val)。

8.2.2 示例分析

【例 8.3】 7 个查询函数的简单应用。

```
//文件名：e8_3.cpp
#include <algorithm>
#include <iostream>
using namespace std;

bool mygreater(int m)
{
    return m>4;
}

int main()
{
    int a[]={1,2,2,2,3,4,4,5,6,7,1,2,2,3};
    int nSize=sizeof(a)/sizeof(int);

    cout<<"原始数组："<<endl;
    for(int i=0; i<nSize; i++)
    {
        cout<<a[i]<<"\t";
    }
    cout<<endl<<endl;
```

```cpp
    int * p1=find(a, a+nSize, 3);
    if(p1!=a+nSize)
        cout<<"(find)首次等于 3 的位置:"<<p1-a<<"\t 值:" << * p1<<endl;

    int * p2=find_if(a, a+nSize, greater);
    if(p2!=a+nSize)
        cout<<"(find_if)首次大于 4 的位置:"<<p2-a<<"\t 值:" << * p2<<endl;

    int b[]={10,12,6};
    int nSize2=sizeof(b)/sizeof(int);
    int * p3=find_first_of(a, a+nSize, b, b+nSize2);
    if(p3!=a+nSize)
        cout<<"(find_first_of)首次在 a 数组中发现 b 数组[10,12,6]中元素位置:"<<p3-a
            <<"\t 值:" << * p3<<endl;

    int * p4=adjacent_find(a, a+nSize);
    if(p4!=a+nSize)
        cout<<"(adjacent_find)首次相邻元素相同位置:"<<p4-a<<"\t 值:" << * p4<<endl;

    int c[]={2,3};
    int nSize3=sizeof(c)/sizeof(int);
    int * p5=find_end(a, a+nSize, c, c+nSize3);
    if(p5!=a+nSize)
        cout <<"最后一次匹配 c 数组[2,3]位置:"<<p5-a<<endl;

    int * p6=search(a, a+nSize, c, c+nSize3);
    if(p6!=a+nSize)
        cout <<"首次匹配 c 数组[2,3]位置:"<<p6-a<<endl;

    int * p7=search_n(a, a+nSize, 3, 2);
    if(p7!=a+nSize)
        cout <<"首次出现 3 个 2 的位置:"<<p7 -a<<endl;
    return 0;
}
```

【例 8.4】 根据学号查询学生信息,且已知学号是关键字。

```cpp
//文件名: e8_4.cpp
#include<algorithm>
#include<iostream>
#include<vector>
#include<string>
using namespace std;
```

```cpp
class Student
{
public:
    int NO;                                     //学号
    string strName;                             //姓名
    Student(int NO, string strName)
    {
        this->NO=NO;
        this->strName=strName;
    }
    bool operator==(int NO)
    {
        return (this->NO ==NO);
    }
};

int main()
{
    vector<Student>v;
    Student s1(101, "张三");
    Student s2(102, "李四");
    v.push_back(s1);
    v.push_back(s2);

    vector<Student >::iterator begin, end, it_find;   //定义 3 个迭代器
    begin=v.begin();                                  //迭代器起始指针
    end=v.end();                                      //迭代器结束指针

    int nFindNO=102;                                  //查询条件是学号=102
    it_find=find(begin, end, nFindNO);                //返回学号等于 102 的迭代器指针

    cout<<"查询学号为"<<nFindNO <<"的信息:"<<endl;
    if (it_find!=end)                                 // 若找到则显示
        cout<<"学号:"<<(*it_find).NO<<"\t"<<"姓名:" << (*it_find).strName<<endl;
    else                                              // 若没有找到
        cout<<"无该学号学生!"<<endl;
    return 0;
}
```

主要理解 Student 类必须重载运算符"=="。当执行 find(begin, end, nFindNO)语句时,[begin,end)间的每一个迭代指针表示的 Student 对象都要与 nFindNO 比较,判断是否相等,因此必须重载 Student 类中的"=="运算符。由于 nFindNO 是整型数,因此重载的"=="运算符参数必须是整型数。

【例 8.5】 已知学生基本属性：学号（整型，关键字），姓名，成绩。现在要求编写相关的功能类，能添加学生对象，并具有下列查询功能：(1)按学号查询；(2)按一组学号查询；(3)按姓名查询；(4)按成绩查询；(5)按成绩范围查询。并编制简单的测试类加以测试。

分析：

(1) 设计思想：采取基本类、集合类设计方法，在集合类中实现添加、查询功能。

(2) 实现查询功能用哪个具体查询函数呢？对于按学号查询，由于学号是关键字，因此用 find 函数即可；对于按一组学号查询，最好用 find_first_of 函数，但该函数一次只能查到一个学号的学生信息，因此一定要采用循环结构才能完成一组学号的查询；对于按姓名查询，采用 find 函数，由于姓名不是关键字，因此也应该采用循环结构；对于按成绩查询，应该用 find 函数，由于成绩不是关键字，因此也应该采用循环结构；按成绩范围查询，应采用 find_if 函数，当然也是循环结构。

实际解决中一定会遇到很多问题，先看一下代码，再做详细说明。设基础类 Student，集合类 StudentCollect，如下所示。

```cpp
//文件名：e8_5.cpp
#include<algorithm>
#include<iostream>
#include<vector>
#include<string>
using namespace std;

const int NO_FIND=1;
const int GRADE_FIND=2;

class Student
{
public:
    int NO;                                         //学号,关键字
    string strName;                                 //姓名
    int grade;                                      //成绩
    static int mark;                                //查询标识
    Student(int NO, string strName,int grade)
    {
        this->NO=NO;
        this->strName=strName;
        this->grade=grade;
    }
    bool operator==(int n)                          //用于学号或成绩查询
    {
```

```cpp
            if(mark==NO_FIND)                          //按学号查询
                return NO==n;
            else                                       //按成绩查询
                return grade==n;
        }

        bool operator==(string name)                   //用于按姓名查询
        {
            return strName.compare(name)==0;
        }
};

int Student::mark=-1;

ostream& operator<<(ostream& os, Student& s)
{
    os<<s.NO<<"\t"<<s.strName<<"\t"<<s.grade;
    return os;
}

class StudFindIf                                       //用于 find_if 函数的类
{
private:
    int low;
    int high;
public:
    StudFindIf(int low, int high)
    {
        this->low=low;
        this->high=high;
    }
    bool operator()(Student & s)
    {
        return s.grade>=low && s.grade <=high;
    }
};

class StudentCollect                                   //学生集合类
{
    vector<Student>vecStud;
public:
```

```cpp
bool Add(Student& s)                           //添加集合元素
{
    vecStud.push_back(s);
    return true;
}

bool FindByNO(int no)                          //按学号查询
{
    Student::mark=NO_FIND;
    vector<Student>::iterator te=find(vecStud.begin(), vecStud.end(), no);
    if(te!=vecStud.end())                      //说明有查询结果
    {
        cout<< * te<<endl;
    }
    else
    {
        cout<<"学号："<<no<<"没有查询记录"<<endl;
    }

    return true;
}
bool FindByNO(int no[], int nSize)             //按学号数组查询
{
    bool bFind=false;
    Student::mark=NO_FIND;
    vector<Student>::iterator te=find_first_of(vecStud.begin(), vecStud.
    end(),no, no+nSize);
    while(te!=vecStud.end())
    {
        bFind=true;
        cout<< * te<<endl;
        te++;
        te=find_first_of(te, vecStud.end(), no, no+nSize);
    }
    if(!bFind)
        cout<<"没有相关记录"<<endl;
    return true;
}
bool FindByName(string name)                   //按姓名查询
{
    bool bFind=false;
    vector<Student>::iterator te=find(vecStud.begin(), vecStud.end(), name);
    while(te!=vecStud.end())
```

```cpp
        {
            bFind=true;
            cout<< * te<<endl;

            te++;
            te=find(te, vecStud.end(), name);
        }
        if(!bFind)
        {
            cout<<"姓名:"<<name<<"没有查询记录"<<endl;
        }
        return true;
    }
    bool FindByGrade(int grade)              //按成绩查询
    {
        Student::mark=GRADE_FIND;
        bool bFind=false;
        vector<Student>::iterator te=find(vecStud.begin(), vecStud.end(), grade);
        while(te!=vecStud.end())
        {
            bFind=true;
            cout<< * te<<endl;

            te++;
            te=find(te, vecStud.end(), grade);
        }
        if(!bFind)
        {
            cout<<"成绩:"<<grade<<"没有查询记录"<<endl;
        }
        return true;
    }
    bool FindByRange(int low, int high)      //按成绩范围查询
    {
        bool bFind=false;

        StudFindIf sf(low, high);
        vector<Student>::iterator te=find_if(vecStud.begin(), vecStud.end(), sf);
        while(te!=vecStud.end())
        {
            bFind=true;
            cout<< * te<<endl;
```

```cpp
            te++;
            te=find_if(te, vecStud.end(), sf);
        }
        return true;
    }
};

int main()                                          //简单测试
{
    Student s1(101, "张三", 50);
    Student s2(102, "李四", 70);
    Student s3(103, "张三", 60);
    Student s4(104, "王五", 50);
    Student s5(105, "王五", 80);

    StudentCollect manage;
    manage.Add(s1); manage.Add(s2);
    manage.Add(s3); manage.Add(s4);
    manage.Add(s5);

    cout<<"按学号查询(102):"<<endl;
    manage.FindByNO(102);                           //返回学号等于102的迭代器指针
    cout<<"按姓名查询(张三):"<<endl;
    manage.FindByName("张三");
    cout<<"按成绩查询(50):"<<endl;
    manage.FindByGrade(50);

    int a[]={101,105,103,107};
    cout<<"按学号组{101,105,103,107}查询:"<<endl;
    manage.FindByNO(a, sizeof(a)/sizeof(int));

    cout<<"按成绩范围[55,70]查询"<<endl;
    manage.FindByRange(55, 70);
    return 0;
}
```

(1) 集合类 StudentCollect 定义了 vector<Student>学生向量,主要有 1 个添加函数及 5 个查询函数。

(2) 在 5 个查询中:按学号查询、按一组学号查询、按姓名查询、按成绩查询都需要重载基础类 Student 中的 operator＝＝函数,只是参数类型不同。按姓名查询需要重载 bool operator＝＝(string s)函数。按学号和成绩查询需要重载 operator＝＝(int m)函数,那么如何确定参数 m 是学号还是成绩呢？本基础类 Student 类中设计了一个整型且是静态的标识变量 mark，通过设置 mark 值可知现在进行何种查询，从而确定 m 是学号还是成绩。

（3）按成绩范围查询稍显特殊，它不需要重载 operator== 函数，用到的函数是 find_if。本示例中采用了函数对象调用技术，编制了相应的类 StudFindIf 函数对象类，通过构造函数 StudFindIf(int low，int high) 传递了上限下限值，通过重载该类中的 operator()(Student& s) 函数完成了真正的比较过程，也可以说应用 find_if 函数结合函数对象应该能定义比较复杂的查询条件，不像其他 STL 提供的系统查询函数那样，一般只需重载基础类中的 operator== 函数就可以了。

8.3 计数

8.3.1 主要函数

STL 主要提供了以下查询函数。
- count()：在序列中统计某个值出现的次数。
- count_if()：在序列中统计与某谓词（表达式）匹配的次数。

各个函数原型如下所示。

① count

原型：

template<class InIt, class T>
　　size_t **count**(InIt first, InIt last,const T& val);

参数说明：
- InIt：输入迭代器，first 表示起始元素的迭代器指针，last 表示结束元素的迭代器指针。
- T：模板参数类型。

该函数返回[first，last)间的元素数目，这些元素满足 *(first+i)=val；。

② count_if

原型：

template<class InIt, class Pred, class Dist>
　　size_t **count_if**(InIt first, InIt last,Pred pr);

参数说明：
- InIt：输入迭代器，first 表示起始元素的迭代器指针，last 表示结束元素的迭代器指针。
- Pred：普通全局函数，返回值是 bool 类型。

该函数是查询[first, last)间迭代器对应元素 *(first+i)的总数,条件是 pr(*(first+i))返回值是 true。

8.3.2 示例分析

【例 8.6】 求数组中有多少个 0。

```cpp
//文件名: e8_6.cpp
#include<algorithm>
#include<iostream>
using namespace std;
int main() {
    int A[]={ 2, 0, 4, 6, 0, 3, 1, -7 };
    const int N=sizeof(A)/sizeof(int);
    cout<<"Number of zeros: "<<count(A, A+N, 0)<<endl;
    return 0;
}
```

这个例子是比较简单的,再看下述例子。

【例 8.7】 查询有多少学生成绩为 80 分。

```cpp
//文件名: e8_7.cpp
#include<algorithm>
#include<iostream>
#include<vector>
using namespace std;
class Student
{
public:
    int NO;                                    //学号
    string strName;                            //姓名
    int grade;                                 //成绩
    Student(int NO, string strName, int grade)
    {
        this->NO=NO;
        this->strName=strName;
        this->grade=grade;
    }
    bool operator==(int grade)
    {
        return this->grade ==grade;
    }
};
```

```
int main() {
    vector<Student>v;
    Student s1(1000, "张三", 80);
    Student s2(1001, "李四", 85);
    Student s3(1002, "王五", 80);
    Student s4(1003, "赵六", 80);
    v.push_back(s1);
    v.push_back(s2);
    v.push_back(s3);
    v.push_back(s4);
    int nCount=count(v.begin(),v.end(),80);
    cout<<"成绩为80分的人数为:"<<nCount<<endl;
    return 0;
}
```

对此例主要应理解必须重载 Student 类的运算符"＝＝"，这是因为当执行 int nCount＝count(v. begin()，v. end()，80)时，要比较[v. begin，v. end)中各 Student 对象是否与 80 相等。

【例 8.8】 查询有多少学生成绩高于 80 分。

```
//文件名：e8_8.cpp
#include<algorithm>
#include<iostream>
#include<vector>
using namespace std;
class Student
{
public:
    int NO;                                    //学号
    string strName;                            //姓名
    int grade;                                 //成绩
    Student(int NO, string strName, int grade)
    {
        this->NO=NO;
        this->strName=strName;
        this->grade=grade;
    }
    bool operator==(int grade)
    {
        return this->grade==grade;
    }
};

class MatchExpress
```

```
    {
        int grade;
    public:
        MatchExpress(int grade)
        {
            this->grade=grade;
        }
        bool operator()(Student& s)
        {
            return s.grade>grade;
        }
    };

    int main() {
        vector<Student>v;
        Student s1(1000, "张三", 80);
        Student s2(1001, "李四", 85);
        Student s3(1002, "王五", 80);
        Student s4(1003, "赵六", 80);
        v.push_back(s1);
        v.push_back(s2);
        v.push_back(s3);
        v.push_back(s4);
        int nCount=count_if(v.begin(), v.end(), MatchExpress(80));
        cout<<"成绩高于 80 分的人数为:"<<nCount<<endl;
        return 0;
    }
```

很明显,定义了一个函数对象类 MatchExpress,重载了 operator()(Student& s)函数。当执行 main 函数中的 count_if 语句时,把第三个参数 MatchExpress(80)作为函数对象,先调用构造函数给 MatchExpress 的成员变量 grade 赋值为 80,然后向量中的每个学生对象作为函数对象的参数调用 operator()(Student& s)函数,判断该学生的成绩是否大于 80。

当然,可在 MatchExpress 类中定义更多的成员变量,形成更复杂的表达式,依据需求分析来确定。

8.4 比较

8.4.1 主要函数

STL 主要提供了以下比较函数。

- equal()：两个序列中的对应元素都相同时为真。
- mismatch()：找出两个序列相异的第一个元素。

各个函数原型如下所示。

① equal

原型：

```
template<class InIt1, class InIt2>
    bool equal(InIt1 first, InIt1 last, InIt2 x);
template<class InIt1, class InIt2, class Pred>
    bool equal(InIt1 first, InIt1 last, InIt2 x, Pred pr);
```

参数说明：
- InIt1：第一个容器的迭代器，first 表示起始元素的迭代器指针，last 表示结束元素的迭代器指针。
- InIt2：第二个容器的迭代器。
- Pred：二元全局函数或函数对象。

第一个原型含义是：对两个输入迭代器而言，若依次有 $*(first+0) = *(x+0)$，$*(first+1) = *(x+1)$，…，$*[first+(last-first-1)] = *[x+(last-first-1)]$，那么这两个容器序列是相等的。

第二个原型与第一个类似，只不过要定义预判定函数 pr($*(first1+N)$，$*(first2+N)$)。

② mismatch

原型：

```
template<class InIt1, class InIt2>
    pair<InIt1, InIt2>mismatch(InIt1 first, InIt1 last, InIt2 x);
template<class InIt1, class InIt2, class Pred>
    pair<InIt1, InIt2>mismatch(InIt1 first, InIt1 last, InIt2 x, Pred pr);
```

参数说明：
- InIt：输入迭代器，first 表示起始元素的迭代器指针，last 表示结束元素的迭代器指针。
- Pred：二元全局函数或函数对象。

第一个原型含义是：对两个迭代器而言，返回第一对元素不相等时的迭代器指针，保存在 pair 对象中。pair 有两个成员变量：first 和 second，分别表示 InIt1 及 InIt2 不相等时的迭代指针。

第二个原型与第一个类似，只不过要定义预判定函数 pr($*(first1+N)$，$*(first2+M)$)。

8.4.2 示例分析

【例 8.9】 比较两个整型数组是否相等。

//文件名：e8_9.cpp

```cpp
#include<algorithm>
#include<iostream>
using namespace std;
int main() {
    int A1[]={ 3, 1, 4, 1, 5, 9, 3 };
    int A2[]={ 3, 1, 4, 2, 8, 5, 7 };
    const int N=sizeof(A1)/sizeof(int);
    cout<<"Result of comparison: "<<equal(A1, A1+N, A2)<<endl;
    return 0;
}
```

【例 8.10】 寻找两个整型数组元素不相等时的元素值。

```cpp
//文件名: e8_10.cpp
#include<algorithm>
#include<iostream>
using namespace std;
int main() {
    int A1[]={ 3, 1, 4, 1, 5, 9, 3 };
    int A2[]={ 3, 1, 4, 2, 8, 5, 7 };
    const int N=sizeof(A1)/sizeof(int);
    pair<int *, int *>result=mismatch(A1, A1+N, A2);
    cout<<"The first mismatch is in position "<<result.first-A1<<endl;
    cout<<"Values are: "<<*(result.first)<<", "<<*(result.second)<<endl;
    return 0;
}
```

主要应理解 pair<int *, int *>result=mismatch(A1, A1＋N, A2),包括：

(1) 由于 A1、A2 均是整型数,因此 pair 模板的两个参数类型是整型指针int *。可分析出：pair 成员变量 first 表示 A1 的迭代指针, second 表示 A2 的迭代指针。

(2) 执行过程是 *(A1+i)与 *(A2+i)比较,若相同,i 增加 1,继续比较,直至不同时 mismatch 返回 pair 对象。

【例 8.11】 查询第一对成绩不相等学生的信息。

```cpp
//文件名: e8_11.cpp
#include<algorithm>
#include<iostream>
#include<vector>
#include<string>
using namespace std;
class Student
{
public:
```

```
        int NO;                                              //学号
        string strName;                                      //姓名
        int grade;                                           //成绩
        Student(int NO, string strName, int grade)
        {
            this->NO=NO;
            this->strName=strName;
            this->grade=grade;
        }
        bool operator==(Student& s)
        {
            return this->grade==s.grade;
        }
};

int main() {
        vector<Student>     v1;
        Student s1(1001, "aaa", 90);
        Student s2(1002, "bbb", 80);
        Student s3(1003, "ccc", 70);
        v1.push_back(s1),v1.push_back(s2),v1.push_back(s3);
        vector<Student>     v2;
        Student s4(1004, "ddd", 90);
        Student s5(1005, "eee", 80);
        Student s6(1006, "fff", 75);
        v2.push_back(s4),v2.push_back(s5),v2.push_back(s6);
        cout<<"查询第一对成绩不相等学生的信息:"<<endl;
        typedef vector<student> ::Herator it;
        pair<it, it> result=mismatch(v1.begin(), v1.end(), v2.begin());

        Student& stu1=*result.first;
        Student& stu2=*result.second;

        cout <<"学号:"<<stu1.NO<<"\t 姓名:"<<stu1.strName<<"\t 成绩:"<<stu1.grade<<endl;
        cout <<"学号:"<<stu2.NO<<"\t 姓名:"<<stu2.strName<<"\t 成绩:"<<stu2.grade<<endl;
        return 0;
}
```

主要理解以下几点：

（1）由于 v1、v2 是 Student 对象的集合，因此 pair 模板的两个参数类型是 Student *。

（2）由于 pair 模板的两个参数类型是 Student *，因此 *result.first 才表示真正的对象。程序中用了引用 Student& stu1=*result.first，当然用 Student stu1=*result.first 也可以，但不如前者好，这是由于后者是对象复制非引用的缘故。

（3）由于涉及比较两个 Student 对象是否相等，因此在类 Student 中必须重载运算符"=="。

第8章 泛型及其他

```
int NO;                           //学号
string stuName;                   //姓名
int grade;                        //成绩
Student(int NO, string stuName, int grade)
{
    this->NO=NO;
    this->stuName=stuName;
    this->grade=grade;
}
bool operator==(Student s)
{
    return(this->grade==s.grade);
}
};

int main()
{
    vector<Student> v1,v2;
    Student s1(1001,"aaa",90);
    Student s2(1002,"bbb",80);
    Student s3(1003,"cc",70);
    v1.push_back(s1);v1.push_back(s2);v1.push_back(s3);
    vector<Student> v2;
    Student s4(1004,"ee",90);
    Student s5(1005,"ee",80);
    Student s6(1006,"rrr",70);
    v2.push_back(s4);v2.push_back(s5);v2.push_back(s6);
    cout<<"查找第一组成绩=的学生的信息是:"<<endl;
    typedef vector<Student>::iterator It;
    pair<It,It> result=search(v1.begin(),v1.end(),v2.begin(),v2.end());
    
    Students stu1=*result.first;
    Students stu2=*result.second;
    
    cout<<"学号:"<<stu1.NO<<"\t姓名:"<<stu1.stuName<<"\t成绩:"<<stu1.grade<<endl;
    cout<<"学号:"<<stu2.NO<<"\t姓名:"<<stu2.stuName<<"\t成绩:"<<stu2.grade<<endl;
    return 0;
}
```

主要理解以下几点:

(1) 由于v1、v2是Student对象的集合,因此pair相关的两个参数类型是Student*。

(2) 由于pair中存放的是多数类型是Student*,因此,result.first了表示中指向对象的指针用了Student& stu1=*result.first;语句用了Student stu1=*result.first也可以,但不建议,这是由于后者将对象复制出的缘故。

(3) 由于要比较两个Student对象是否相等,因此在类Student中必须重载运算符

第9章 变异算法

变异算法即变异函数,其主要特点是修改容器中的元素。如修改容器中的元素值,改变容器中的元素序列等。表 9.1 所示为按照功能划分变异函数。

表 9.1 变异函数列表

序号	功能	函数名称	说明
1	复制	copy	从序列的第一个元素起进行正向复制
		copy_backward	从序列的最后一个元素起进行反向复制
2	交换	swap	交换两个元素
		swap_ranges	交换指定范围的元素
		iter_swap	交换由迭代器所指的两个元素
3	变换	transform	将某操作应用于指定范围的每个元素
4	替换	replace	用一个给定值替换一些值
		replace_if	替换满足谓词的一些元素
		replace_copy	复制序列时用一给定值替换元素
		replace_copy_if	复制序列时替换满足谓词的元素
5	填充	fill	用一给定值填充所有元素
		fill_n	用一给定值填充前 n 个元素
6	生成	generate	用一操作的结果填充所有元素
		generate_n	用一操作的结果填充前 n 个元素
7	删除	remove	删除具有给定值的元素
		remove_if	删除满足谓词的元素
		remove_copy	复制序列时删除具有给定值的元素
		remove_copy_if	复制序列时删除满足谓词的元素
8	唯一	unique	删除相邻的重复元素
		unique_copy	复制序列时删除相邻的重复元素

续表

序号	功能	函数名称	说明
9	反转	reverse	反转元素的次序
		reverse_copy	复制序列时反转元素的次序
10	环移	rotate	循环移动元素
		rotate_copy	复制序列时循环移动元素
11	随机	random_shuffle	采用均匀分布来随机移动元素
12	划分	partition	将满足某谓词的元素都放到前面
		stable_partition	将满足某谓词的元素都放到前面并维持原顺序

9.1 复制

9.1.1 主要函数

复制是应用很广的一个功能，主要函数有：
- copy()：从序列的第一个元素起进行正向复制。
- copy_backward()：从序列的最后一个元素起进行反向复制。

各个函数的原型如下所示。

① copy

原型：

```
template<class InIt, class OutIt>
OutIt copy(InIt first, InIt last, OutIt x);
```

参数说明：
- 模板参数 InIt 表示输入迭代器，OutIt 表示输出迭代器。
- 返回值类型是 OutIt，是输出迭代器的尾指针。

该函数功能是"正向—正向"复制，把输入迭代器[first, last)间的元素依次复制到输出迭代器 x 中，并返回输出迭代器的尾指针。

② copy_backward

原型：

```
template<class BidIt1, class BidIt2>
BidIt2 copy_backward(BidIt1 first, BidIt1 last, BidIt2 x);
```

参数说明：
- 模板参数 BidIt1，BidIt2 是双向迭代器。前者功能是读，后者是写。

- 返回值类型是 BitIt2,是写功能输出迭代器的首指针。

该函数功能是"反向—反向"复制,把 BidIt1 中的(last,first]间元素依次复制到写双向迭代器 x 中,返回写功能输出迭代器的首指针。

9.1.2 示例分析

【例 9.1】 copy 简单示例。

```
//文件名:e9_1.cpp
#include<iostream>
#include<vector>
#include<iterator>
using namespace std;
int main()
{
    int a[5]={1,2,3,4,5};
    int b[5];
    vector<int>v;

    copy(a, a+5, b);                                      //a 数组复制到 b 数组
    copy(a, a+5, back_inserter(v));                       //a 数组复制到向量 vector 中

    cout<<"原始 a 数组为:";                                //显示原始数组
    copy(a, a+5, ostream_iterator<int>(cout, "\t")); cout<<endl;    //1 2 3 4 5
    cout<<"b 数组为:";                                     //显示复制后的 b 数组
    copy(b, b+5, ostream_iterator<int>(cout, "\t")); cout<<endl;    //1 2 3 4 5
    cout<<"vector 向量为:";                                //显示复制后的向量
    copy(v.begin(), v.end(), ostream_iterator<int>(cout, "\t")); cout<<endl;
                                                          //1 2 3 4 5
    return 0;
}
```

简单分析如下:

(1) copy 函数非常简洁,以前编程往往用一个循环来完成源数据到目的数据的复制,现在仅用一条语句,只要给出源数据待复制区间的迭起止指针及输出迭代器起始指针就可以了。但一定要注意,根据说明[first,last),左边是闭区间,右边是开区间。所以 a 数组全部复制到 b 数组一定是 copy(a,a+5,b),而不能是 copy(a,a+4,b)。

(2) 要注意:若目的数据类型是数组,则一定要保证它的内存空间大于或等于源数据待复制空间的大小。例如本题中是 b[5],与源复制数据 a[5]内存空间大小相等;若目的数据类型是基本序列容器 vector、list 等,则由于可通过 back_inserter,front_inserter 插入迭代器动态改变容器的大小,因此对基本序列容器对象内存空间没有特殊限制。但是若不用

back_inserter，例如本题直接写成 copy(a，a+5，v)，则编译通不过。

(3) copy 语义非常丰富。例如本题中可将数组复制到数组、数组复制到基本序列容器、数组复制到屏幕、基本序列容器数据复制到屏幕等。这其中，我们发现 STL 预定义迭代器 ostream_iterator 及 istream_iterator 起着很重要的作用。再举一个例子，若把键盘输入的整型数存放到某 vector 对象 vec 中，可写为 copy(istream<int>(cin)，istream_iterator<int>()，back_inserter(vec))。

9.2 交换

9.2.1 主要函数

主要函数有：
- swap()：交换两个元素。
- swap_ranges()：交换指定范围的元素。
- iter_swap()：交换由迭代器所指的两个元素。

各个函数的原型如下所示。

① swap

原型：

template<class T>
void **swap**(T & a,T & b);

两个相同模板参数类型变量的值互相交换。

② swap_ranges

原型：

template<class FwdIt1, class FwdIt2>
FwdIt2 **swap_ranges**(FwdIt1 first, FwdIt1 last, FwdIt2 x);

参数说明：

FwdIt：表明模板参数是前向迭代器。

该函数的功能是迭代器[first，last)间表示的容器 A 中元素与容器 B 中迭代器为 x 处开始的元素依次交换，并返回容器 B 中最后交换元素的迭代器指针。

③ iter_swap

原型：

template<class FwdIt1, class FwdIt2>
void iter_swap(FwdIt1 x, FwdIt2 y);

参数说明：

FwdIt：表明模板参数是前向迭代器。

该函数的功能是交换两个前向迭代指针指向的元素值。

9.2.2 示例分析

【例 9.2】 交换简单示例。

```cpp
//文件名：e9_2.cpp
#include<iostream>
#include<vector>
#include<iterator>
#include<algorithm>
using namespace std;
int main()
{
    int a=10;
    int b=20;
    cout<<"原始数据 a="<<a<<"\tb="<<b<<endl;
    swap(a, b);                                                  //交换两个数据
    cout<<"交换后数据 a="<<a<<"\tb="<<b<<endl<<endl;

    int a2[5]={1,2,3,4,5};                                       //交换两个数组元素
    int b2[5]={6,7,8,9,10};
    cout<<"原始 a2[5]=";
    copy(a2, a2+5, ostream_iterator<int>(cout, "\t"));
    cout<<endl;
    cout<<"原始 b2[5]=";
    copy(b2, b2+5, ostream_iterator<int>(cout, "\t"));
    cout<<endl;
    swap_ranges(a2, a2+5, b2);
    cout<<"交换后 a2[5]=";
    copy(a2, a2+5, ostream_iterator<int>(cout, "\t"));
    cout<<endl;
    cout<<"交换后 b2[5]=";
    copy(b2, b2+5, ostream_iterator<int>(cout, "\t"));
    cout<<endl<<endl;

    int a3[5]={10,20,30,40,50};
    int b3[5]={15,25,35,45,55};
    vector<int>v1(a3, a3+5);                                     //交换两个基本序列元素
    vector<int>v2(b3, b3+5);
    cout<<"原始 vector1=";
    copy(v1.begin(), v1.end(), ostream_iterator<int>(cout, "\t"));
```

```
            cout<<endl;
            cout<<"原始 vector2=";
            copy(v2.begin(), v2.end(), ostream_iterator<int>(cout, "\t"));
            cout<<endl;
            swap(v1, v2);
            cout<<"交换后 v1=";
            copy(v1.begin(), v1.end(), ostream_iterator<int>(cout, "\t"));
            cout<<endl;
            cout<<"交换后 vector2=";
            copy(v2.begin(), v2.end(), ostream_iterator<int>(cout, "\t"));
            cout<<endl;
            return 0;
        }
```

(1) 可以看出交换函数可应用在基本数据类型之间、数组之间、基本序列容器之间。

(2) 同学们可能会问：这三个函数到底有什么区别？哪些能互相代替？其实，通过 swap 函数原型 swap(T & a, T & b)可明确看出它相当于引用调用，其余两个函数 swap_ranges、iter_swap 相当于地址指针调用。因此示例中：基本数据类型 swap(a,b)可以用 iter_swap(&a, &b)代替；基本序列容器 swap(v1, v2)可以用 iter_swap(&v1, &v2)代替，当然从功能上说也可用 swap_ranges(v1.begin(), v1.end(), v2.begin())代替；但是对数组来说只能用 swap_ranges，如示例那样。因此可以总结出：对于基本数据类型可用 swap 或 iter_swap，对数组只能用 swap_ranges，对基本序列容器三个函数均可。

(3) 对数组而言，要交换的必须是真实的内存空间，因为对定义好的数组而言，它的大小是固定的，不可能动态变化，因此 int a[5]全部数据不能与 int b[7]的全部数据交换；但是对基本序列容器而言，由于它支持元素空间动态分配，因此相交换的容器拥有的元素内存空间可以是不同的，如下代码是正确的：

```
int a[5]={1,2,3,4,5};
int b[3]={6,7};
vector<int>v1(a, a+5);
vector<int>v2(b, b+2);
swap(v1, v2);
```

v1 有 5 个元素，v2 有 2 个元素，交换后 v1 有 2 个元素，v2 有 5 个元素。

9.3 变换

9.3.1 主要函数

主要函数有：
transform()：将某操作应用于指定范围的每个元素。

函数原型如下所示。

```
template<class InIt, class OutIt, class Unop>
    OutIt transform(InIt first, InIt last, OutIt x, Unop uop);
template<class InIt1, class InIt2, class OutIt, class Binop>
    OutIt transform(InIt1 first1, InIt1 last1, InIt2 first2,OutIt x, Binop bop);
```

参数说明：
- InIt、OutIt：输入输出迭代器。
- Unop：一元函数。
- Binop：二元函数。

第一个模板函数功能是：一个输入迭代器容器每个元素依次作为一元函数 uop 参数传入并执行，结果输出到输出迭代器 x 表示的容器中。即 $*(x+N) = uop(*(first+N))$，$N \in [0, last-first)$。

第二个模板函数功能是：两个输入迭代器容器对应的一对元素作为二元函数 bop 参数传入并执行，结果输出到输出迭代器 x 表示的容器中。即 $*(x+N) = bop(*(first1+N)$，$*(first2+N))$，$N \in [0, last1-first1)$。

9.3.2　示例分析

【例 9.3】　变换简单示例。

```cpp
//文件名：e9_3.cpp
#include<iostream>
#include<vector>
#include<iterator>
#include<algorithm>
using namespace std;
int func1(int value)
{
    return value * 2;
}
int func2(int value1, int value2)
{
    return value1+value2;
}

int main()
{
    int a[5]={1,2,3,4,5};
    vector<int>v1(a, a+5);
    vector<int>v2(5);          //v2 初始容器大小有 5 个 int 元素
```

```cpp
        cout<<"原始向量 v1=";
        copy(v1.begin(), v1.end(), ostream_iterator<int>(cout, "\t"));
        cout<<endl;
        cout<<"v1 * 2-->v1=";
        transform(v1.begin(), v1.end(), v1.begin(), func1);
                                //v1元素调用一元函数,并修改自己:func1(v1)-->v1
        copy(v1.begin(), v1.end(), ostream_iterator<int>(cout, "\t"));
        cout<<endl;
        cout<<"v1 * 2-->v2=";
        transform(v1.begin(), v1.end(), v2.begin(), func1);
                                //v1元素调用一元函数,结果给 v2:func1(v1)-->v2
        copy(v2.begin(), v2.end(), ostream_iterator<int>(cout, "\t"));
        cout<<endl<<endl;

        int a2[5]={1,2,3,4,5};
        int b2[5]={6,7,8,9,10};
        int c2[5];
        cout<<"a2[5]=";
        copy(a2, a2+5, ostream_iterator<int>(cout, "\t"));
        cout<<endl;
        cout<<"b2[5]=";
        copy(b2, b2+5, ostream_iterator<int>(cout, "\t"));
        cout<<endl;
        cout<<"a2+b2-->c2=";
        transform(a2, a2+5, b2, c2, func2);
                                //a2,b2元素调用二元函数,结果给 c2:func2(a2,b2)-->c2
        copy(c2, c2+5, ostream_iterator<int>(cout, "\t"));
        return 0;
}
```

(1) 演示了 transform 如何调用一元函数及二元函数。结果可以修改自己,如题中 transform(v1.begin(), v1.end(), v1.begin(), func1);也可以形成新的集合,如题中 transform(v1.begin(), v1.end(), v2.begin(), func1)。

(2) 为了说明问题,从示例中抽取出如下几行代码:

```cpp
vector<int>v1(a, a+5);
vector<int>v2(5);
transform(v1.begin(), v1.end(), v2.begin(), func1);
```

特别注意第 2 行 vector<int> v2(5),即 v2 初始容器大小有 5 个 int 元素。如果把这行修改为 vector<int> v2,其余不变,就会发现:编译通过,但执行会出错误。那么,是否说明结果容器大小是否必须大于或等于 v1 容器大小呢? 不一定,通过加 back_inserter 就

可以解决这个问题,如下:

```
vector<int>v1(a, a+5);
vector<int>v2;
transform(v1.begin(), v1.end(), back_inserter(v), func1);
```

(3) 既然调用的是一元或二元函数,而且题中的功能比较简单,都可以应用 STL 中提供的系统函数对象。如题中 transform(v1.begin(), v1.end(), v2.begin(), func1)可以修改为 transform(v1.begin(), v1.end(), v2.begin(), bind2nd(multiplies<int>(), 2));题中 transform(a2, a2+5, b2, c2, func2)可以修改为 transform(a2, a2+5, b2, c2, plus<int>())。当然,不要忘记加包含文件 #include <functional>。

【例 9.4】 字符串加密后输出。要求:在文本文件中存有明文,以行为单位加密后输出至屏幕。

```cpp
//文件名:e9_4.cpp
#include<iostream>
#include<fstream>
#include<string>
#include<vector>
#include<iterator>
#include<algorithm>
#include<functional>
using namespace std;

template<class T>
class Encrypt
{
};

template<>
class Encrypt<string>                    //模板特化
{
public:
    string operator()(const string& src)  //对字符串加密
    {
        string s=src;
        int len=s.length();
        for(string::iterator it=s.begin(); it!=s.end(); it++)
        {
            *it= *it+1;                   //加密算法:对应字符 ASCII 加 1
        }
        return s;
    }
};
```

```
int main()
{
    string strText;
    vector<string>v;
    vector<string>vResult;
    ifstream in("d:\\data.txt");
    while(!in.eof())                                      //读文件
    {
        getline(in, strText, '\n');                       //读一行
        v.push_back(strText);
    }
    in.close();
    transform(v.begin(), v.end(), back_inserter(vResult), Encrypt<string>());
                                                          //加密变换
    copy(vResult.begin(), vResult.end(), ostream_iterator<string>(cout, "\n"));
                                                          //输出至屏幕
    return 0;
}
```

(1) 可以看出一元函数 Encrypt 采取了模板特化技术，在 class Encrypt<string>中封装了对字符串的加密操作。模板特化非常易于程序的扩充，如果现在增加一个对整型数据的加密，只需加一个模板特化类即可：

```
template<>
    class Encrypt<int>                                    //模板特化
    {
    public:
        int operator()(const int& src)                    //对整型数加密算法
        {
            ...
        }
    };
```

(2) 函数对象类 Encrypt<string>非常简单，字符串中对应字符 ASCII 加 1 即是加密结果。

9.4 替换

9.4.1 主要函数

主要函数有：

- replace()：用一个给定值替换一些值。

- replace_if()：替换满足谓词的一些元素。
- replace_copy()：复制序列时用一给定值替换元素。
- replace_copy_if()：复制序列时替换满足谓词的元素。

函数原型如下所示。

① replace

原型：

template<class FwdIt, class T>
void **replace**(FwdIt first, FwdIt last,const T& vold, const T& vnew);

参数说明：
- FwdIt：前向迭代器。
- T：容器元素类型。

该函数功能是：遍历容器序列，若某元素等于旧值，则用新值代替。即：

if(*(first+N)==vold)   N∈[0,last-first)
    *(first+N)=vnew

② replace_if

原型：

template<class FwdIt, class Pred, class T>
void **replace_if**(FwdIt first, FwdIt last,Pred pr, const T & val);

参数说明：
- FwdIt：前向迭代器。
- Pred：一元判定函数。
- T：容器元素类型。

该函数功能是：

if(pr(*(first+N)))   N∈[0,last-first)
*(first+N)=val;

③ replace_copy

原型：

template<class InIt, class OutIt, class T>
OutIt **replace_copy**(InIt first, InIt last, OutIt x,const T & vold, const T & vnew);

参数说明：
- Init：输入迭代器。
- OutIt：输出迭代器。
- T：容器元素类型。

该函数功能是：

```
if (*(first+N)==vold)    N∈[0,last-first)
        *(x+N)=vnew;
else
        *(x+N)=*(first+N)
```

④ replace_copy_if

原型：

```
template<class InIt, class OutIt, class Pred, class T>
OutIt replace_copy_if(InIt first, InIt last, OutIt x, Pred pr, const T& val);
```

参数说明：
- Init：输入迭代器。
- OutIt：输出迭代器。
- Pred：一元判定函数。
- T：容器元素类型。

该函数功能是：

```
if (pr(*(first+N)))      N∈[0,last-first)
        *(x+N)=val;
else
        *(x+N)=*(first+N)
```

9.4.2 示例分析

【例 9.5】 4 个 replace 函数简单示例。

```cpp
//文件名：e9_5.cpp
#include<iostream>
#include<vector>
#include<iterator>
#include<algorithm>
#include<functional>
using namespace std;
int main()
{
    int a[9]={1,2,3,4,5,4,3,2,1};
    cout<<"原始数据：";
    copy(a, a+9, ostream_iterator<int>(cout, "\t"));
    cout<<endl;                                          //1 2 3 4 5 4 3 2 1

    cout<<"原数据 2 用 10 代替(replace)：";
```

```
        vector<int>v1(a, a+9);
        replace(v1.begin(), v1.end(), 2,10);
        copy(v1.begin(), v1.end(), ostream_iterator<int>(cout, "\t"));    //1 10 3 4 5 4 3 10 1
        cout<<endl;

        cout<<"原数据<4 的用 20 代替(replace_if):";
        vector<int>v2(a, a+9);
        replace_if(v2.begin(), v2.end(), bind2nd(less<int>(), 4), 20);
        copy(v2.begin(), v2.end(), ostream_iterator<int>(cout, "\t"));
                                                                //20 20 20 4 5 4 20 20 20
        cout<<endl;

        cout<<"原数据 4 用 30 代替,且 v3->v4(replace_copy):";
        vector<int>v3(a, a+9);
        vector<int>v4;
        replace_copy(v3.begin(), v3.end(), back_inserter(v4), 4, 30);    //1 2 3 30 5 30 3 2 1
        copy(v4.begin(), v4.end(), ostream_iterator<int>(cout, "\t"));
        cout<<endl;

        cout<<"原数据<4 用 40 代替,且 v5->v6(replace_copy_if):";
        vector<int>v5(a, a+9);
        vector<int>v6;
        replace_copy_if(v5.begin(), v5.end(), back_inserter(v6), bind2nd(less<int>(), 4), 40);
        copy(v6.begin(), v6.end(), ostream_iterator<int>(cout, "\t"));
                                                                //40 40 40 4 5 4 40 40 40
        cout<<endl;
        return 0;
}
```

【例 9.6】 一个 replace 易犯错误示例。

```
//文件名:e9_6.cpp
#include <iostream>
#include <algorithm>
#include <iterator>
using namespace std;
int main() {
        int a[]={2,1,3,2,2,5};
        replace(a, a+6, a[0], 10);
        copy(a, a+6, ostream_iterator<int>(cout, "\t"));         //10 1 3 2 2 5
        return 0;
}
```

这看起来是一个非常简单的例子，完成的功能是把数组 a[] 中所有整型数 2 用 10 来代替。但结果却是"10 1 3 2 2 5"，仅第一个 2 变成了 10，这是什么原因呢？看一下 replace 的定义，就很容易发现原因：

void **replace**(FwdIt first, FwdIt last, const T & vold, const T & vnew)

vold 参数是引用调用，而非传值调用。对本例而言，初始时 vold=2，按表意形式相当于 replace(a, a+6, 2, 10)，但当第一次发现数组中元素 2，恰好是 a[0]，这样 a[0] 就变成 10。由于是引用调用，相当于 vold=10，这样当继续扫描数组 a 时，按表意形式相当于 replace(a+1, a+6, 10, 10)，因为 vold=vnew=10，所以 a 数组后续元素不可能再发生变化。

那么，如何完成所需功能呢？参考 replace 函数定义，只要保证 vold 内存空间不在 *[first, last) 内即可。对本题而言，修改如下：

int vold=a[0];
replace(a, a+6, vold, 10)

vold 与 a[0] 值相等，但不在同一内存空间。

9.5 填充

9.5.1 主要函数

主要函数有：
- fill()：用一给定值填充所有元素。
- fill_n()：用一给定值填充前 n 个元素。

函数原型如下所示。

① fill

原型：

template<class FwdIt, class T>
void **fill**(FwdIt first, FwdIt last, const T & x);

参数说明：
- FwdIt：前向迭代器。
- T：容器元素类型。

该函数功能是：遍历容器中所有元素，每个元素都赋成值 x。即：

*(first+N)=x, N∈[0, last-first)

② fill_n

原型：

```
template<class OutIt, class Size, class T>
void fill_n(OutIt first, Size n, const T & x);
```

参数说明:
- OutIt: 输出迭代器。
- Size: 要填充的元素个数。
- T: 容器元素类型。

该函数功能是: 遍历容器 Size 个元素, 每个元素都赋成值 x。即:

* (first+N)=x, N∈ [0,n)

9.5.2 示例分析

【例 9.7】 fill 函数简单示例。

```
#include<iostream>
#include<vector>
#include<iterator>
#include<algorithm>
using namespace std;
int main()
{
    int a[5];
    fill(a, a+5, 0);
    cout<<"a[5]=";
    copy(a, a+5, ostream_iterator<int>(cout, "\t"));
    cout<<endl;

    vector<int>v1(5);
    fill(v1.begin(), v1.end(), 10);
    cout<<"vector v1=";
    copy(v1.begin(), v1.end(), ostream_iterator<int>(cout, "\t"));  //10 10 10 10 10
    cout<<endl;

    vector<int>v2;
    fill_n(back_inserter(v2), 5, 20);
    cout<<"vector v2=";
    copy(v2.begin(), v2.end(), ostream_iterator<int>(cout, "\t"));  //20 20 20 20 20
    cout<<endl;
    return 0;
}
```

运用 fill 函数填充,一定要注意迭代器的有效性。如 vector<int> v1(5),说明 v1 初始有 5 个整型元素空间,所以 fill(v1.begin(), v1.end(), 10)才是有效的,给 5 个元素都赋值 10。如果写成 vector<int> v1,则由于[v1.begin(), v1.end())间没有元素,fill (v1.begin(),v1.end(),10)也是无意义的。

因为没有给定第二个 vector<int> v2 容器的初始大小,因此必须使用 back_inserter() 来添加新元素,当然它以下面代码也是等价的。

```
vector<int>v3(5);
fill(v3.begin(), 5, 20);
```

9.6 生成

9.6.1 主要函数

主要函数有:
- generate():用一操作的结果填充所有元素。
- generate_n():用一操作的结果填充前 n 个元素。

函数原型如下所示。

① generate

原型:

template<class FwdIt, class Gen>
void **generate**(FwdIt first, FwdIt last, Gen g);

参数说明:
- FwdIt:前向迭代器。
- Gen:函数对象。

该函数功能是:

*(first+N)=g(); N∈[0,last-first)

② generate_n

原型:

template<class OutIt, class Pred, class Gen>
void **generate_n**(OutIt first, Dist n, Gen g);

参数说明:
- OutIt:输出迭代器。
- Gen:函数对象。

该函数功能是:

*(first+ N)= g(); N∈[0,n)

9.6.2 示例分析

【例 9.8】 生成斐波那契数列。

```cpp
//文件名：e9_8.cpp
#include<iostream>
#include<vector>
#include<algorithm>
#include<iterator>
using namespace std;
int Fibonacci()
{
    static int r;
    static int f1=0;
    static int f2=1;
    r=f1+f2;
    f1=f2;
    f2=r;
    return f1;
}
int main() {
    vector<int>v1(10);
    generate(v1.begin(),v1.end(),Fibonacci);
    cout<<"0,1开始前10个斐波那契数列为："<<endl;
    copy(v1.begin(),v1.end(),ostream_iterator<int>(cout,"\t"));
    cout<<endl;
    return 0;
}
```

生成斐波那契数列对本示例而言应该是比较简单的，但马上就会发现：如果想再重新生成相同的斐波那契数列是不可能的，这是由于 Fibonacci 函数中 f1,f2 是局部静态变量，按照本算法，它们必定是递增的。因此，若想再重新生成斐波那契数列，一定要重新给 f1,f2 赋初值。一个办法是把 f1,f2 变量声明成全局变量，就可以在任意位置对其赋初值了。当然这个方法可行，但不是最优的，最好的办法是把它封装成如下的函数对象。

```cpp
class Fibonacci
{
    int f1;
    int f2;
public:
    Fibonacci(int start1, int start2)
    {
```

```
        f1=start1;
        f2=start2;
    }
    int operator()()
    {
        int r=f1+f2;
        f1=f2;
        f2=r;
        return r;
    }
};
```

例如调用如下代码,v1,v2 容器有相同排列的斐波那契数列。

```
vector v1(10);
generate(v1.begin(), v1.end(), Fibonacci(0,1));
copy(v1.begin(), v1.end(), ostream_iterator<int>(cout, "\t"));
vector v2(10);
generate(v2.begin(), v2.end(), Fibonacci(0,1));
copy(v1.begin(), v1.end(), ostream_iterator<int>(cout, "\t"));
```

可以看出:v2 之所以与 v1 有相同的排列,关键在于 Fibonacci(0,1)调用构造函数一方面重新生成了新的 Fibonacci 对象,一方面又给斐波那契数列初值 f1,f2 重新赋值。利用函数对象可以封装许多初始化信息,隐蔽性、封装性更好,因此优于全局函数。

【例 9.9】 生成随机数。要求产生 10 个[0,100]间整型数,产生 10 个[0,1)间小数。

分析:(1)最好编制一个产生随机数的框架。一方面产生整型及浮点随机数的算法是不同的,另一方面希望它们有共同接口。能较好地符合这两方面的是模板特化技术。(2)产生随机数最好每次起始"种子"都不同,这样产生的数才更"随机",那么什么能保证"种子"变化呢?一个很好的方案是采用"时间函数"做随机数种子。依据这两条,编制代码如下所示。

```
//文件名:e9_9.cpp
#include<iostream>
#include<vector>
#include<algorithm>
#include<iterator>
#include<time.h>
using namespace std;

template<class T>
class MyRandom
{
};
```

```cpp
template<>
class MyRandom<int>
{
public:
    MyRandom()
    {
        srand(time(NULL));
    }
    int operator()()
    {
        int result=rand()%100;
        return result;
    }
};
template<>
class MyRandom<float>
{
public:
    MyRandom()
    {
        srand(time(NULL));
    }
    float operator()()
    {
        float result=rand()%100*0.01f;
        return result;
    }
};

int main()
{
    cout<<"产生[0,100)间 10个整型随机数为："<<endl;
    vector<int>v1(10);
    generate_n(v1.begin(), 10, MyRandom<int>());
    copy(v1.begin(), v1.end(), ostream_iterator<int>(cout, "\t"));
    cout<<endl;

    cout<<"产生[0,1)间 10个浮点随机数为："<<endl;
    vector<float>v2(10);
    generate_n(v2.begin(), 10, MyRandom<float>());
    copy(v2.begin(), v2.end(), ostream_iterator<float>(cout, "\t"));
    cout<<endl;
    return 0;
}
```

(1) 随机数泛型类 MyRandom<T>,两个特化类 MyRandom<int>及 MyRandom<float>。很明显它们的接口形式是相同的。如果需要增加新数据类型的随机数,只须增加一个相应的模板特化类就可以了。

(2) 在特化类构造函数中都执行了 srand(time(NULL)),这样每次产生新的 MyRandom 函数对象时,由于时间的不同,也就决定了每次产生的随机数种子不同,进而决定产生的随机数序列不同。由于用到了系统时间及随机数函数,因此要包含文件♯include <time.h>及♯include <stdlib.h>。

(3) 产生整型随机数算法:系统 rand()返回一个随机非负整数,若保证在[0,100)区间,只须对 100 取余即可。产生浮点随机数算法:先通过 rand()％100 得到一个[0,100)区间整数,再缩小 100 倍,即乘以 0.01,得到小数在[0,1)区间。

(4) 本例产生随机数在[0,100)及[0,1)区间,那么如果产生任意要求范围内的整型数或浮点数,应该如何修改源程序呢?同学们可以思考一下。

【例 9.10】 保存二次曲线的坐标点。例如圆,从 0~360°每间隔已知角度记录圆上点的坐标。

分析:(1)每个点有横坐标及纵坐标,因此要有描述基本点的结构体 Point;(2)二次曲线有很多,如圆、双曲线和抛物线等。尽管获得基本点的算法不同,但是希望统一接口,因此要用到模板特化技术。代码如下所示。

```
//文件名:e9_10.cpp
#include<iostream>
#include<vector>
#include<algorithm>
#include<math.h>
#include<iterator>
using namespace std;
struct Point
{
    float x;
    float y;
};
class CircleTag
{
};
template<class T>
class MyCurve
{
};
template<>
class MyCurve<CircleTag>
```

```
        float ox;                       //圆心 x 坐标
        float oy;                       //圆心 y 坐标
        float r;                        //半径
        int angle;                      //每间隔多少角度取一个圆上坐标
    public:
        MyCurve(float ox, float oy, float r, int angle)
        {
            this->ox=ox, this->oy=oy, this->r=r;
            this->angle=angle;
        }
        Point operator()()
        {
            Point pt;
            static int curAngle=0;
            pt.x=ox+r * cos(curAngle/360.0f * 2 * 3.14f);
            pt.y=oy+r * sin(curAngle/360.0f * 2 * 3.14f);

            curAngle+=angle;

            return pt;
        }
};
ostream & operator<< (ostream & os, const Point & t)     //为了可以用 cout<<t
{
    os<<"("<<t.x<<","<<t.y<<")";
    return os;
}
int main(void)
{
    vector<Point>v(10);
    generate(v.begin(), v.end(), MyCurve<CircleTag>(10.0f,10.0f,10.0f,36));
                                                         //圆心(10,10),半径 10
    cout<<"从 0 到 360 度每转 36 度取一点坐标为:"<<endl;   //每转 36°取圆上一点坐标
    copy(v.begin(), v.end(), ostream_iterator<Point>(cout, "\t"));
    return 0;
}
```

（1）二次曲线泛型类 MyCurve<T>,模板特化类 MyCurve<CircleTag>。如果再增加一个新的二次曲线,只要再增加一个模板特化类就可以了。

（2）同学们会发现有一个 CircleTag 类,内容为空,主函数中用到的地方是 generate 第三个参数,为 MyCurve<CircleTag>(10.0f,10.0f,10.0f,36)。其实 CircleTag 类在本例中

只是起一个标识作用,当调用时看到有 MyCurve<CircleTag>,通过尖括号中的内容就会知道要求的是圆的坐标。可以反过来想,如果没有 CircleTag 类,那么调用时统一接口 MyCurve<>的尖括号中写什么内容加以区分是圆还是双曲线呢?

(3) 获得圆上点坐标是依据参数方程 $x=x0+r\cos\theta, y=y0+r\sin\theta$ 得到的。

(4) 为了利用标准输出流直接输出 Point 结构体的内容,必须重载 ostream & operator<<(ostream & os, const Point & t)。

9.7 删除

9.7.1 主要函数

主要函数有:
- remove():删除具有给定值的元素。
- remove_if():删除满足谓词的元素。
- remove_copy():复制序列时删除具有给定值的元素。
- remove_copy_if():复制序列时删除满足谓词的元素。

函数原型如下所示。

① remove

原型:

```
template<class FwdIt, class T>
FwdIt remove(FwdIt first, FwdIt last, const T& val);
```

参数说明:
- FwdIt:前向迭代器。
- T:容器元素类型。

该函数功能是:首先设变量 X=first,最终返回值是 X。

```
if (!(*(first+N)==val))    N∈[0,last-first)
    *X++=*(first+N);
```

② remove_if

原型:

```
template<class FwdIt, class Pred>
FwdIt remove_if(FwdIt first, FwdIt last, Pred pr);
```

参数说明:
- FwdIt:前向迭代器。
- Pred:一元判定函数。

该函数功能是：首先设变量 X=first，最终返回值是 X。

```
if (!pr(*(first+N)))         N∈[0,last-first)
    *X++=*(first+N);
```

③ remove_copy

原型：

template<class InIt, class OutIt, class T>
OutIt **remove_copy**(InIt first, InIt last, OutIt x, const T& val);

参数说明：
- InIt：输入迭代器。
- OutIt：输出迭代器。
- T：值类型。

该函数功能是：

```
if (!(*(first+N)==val))      N∈[0,last-first)
*x++=*(first+N);
```

④ remove_copy_if

原型：

template<class InIt, class OutIt, class Pred>
OutIt **remove_copy_if**(InIt first, InIt last, OutIt x, Pred pr);

参数说明：
- InIt：输入迭代器。
- OutIt：输出迭代器。
- Pred：一元判定函数。

该函数功能是：

```
if (!pr(*(first+N)))         N∈[0,last-first)
*x++=*(first+N);
```

9.7.2 示例分析

【例 9.11】 remove 简单示例。

```
//文件名：e9_11.cpp
#include<iostream>
#include<vector>
#include<algorithm>
#include<functional>
#include<iterator>
```

```cpp
using namespace std;
int main(void)
{
    int a[]={1,2,2,4,5,4,3,2,1};
    vector<int>v1(a, a+9);

    cout<<"删除前向量 v1=:";
    copy(v1.begin(), v1.end(), ostream_iterator<int>(cout, "\t"));   //1 2 2 4 5 4 3 2 1
    cout<<endl;

    vector<int>::iterator first=v1.begin();
    vector<int>::iterator last=NULL;
    last=remove(v1.begin(), v1.end(), 2);                            //删除等于 2 元素

    cout<<"删除后向量 v1=";
    copy(v1.begin(), v1.end(), ostream_iterator<int>(cout, "\t"));   //1 4 5 4 3 1 3 2 1
    cout<<endl;
    cout<<"删除后有效数据=";
    copy(first, last, ostream_iterator<int>(cout, "\t"));            //1 4 5 4 3 1
    cout<<endl;
    return 0;
}
```

执行结果为：

删除前向量 v1=1 2 2 4 5 4 3 2 1
删除后向量 v1=1 4 5 4 3 1 3 2 1
删除后有效数据=1 4 5 4 3 1

经过比较发现，remove 前后容器的内存空间没有变化，前后均是 9 个整型元素空间，也就是说 remove 并不是删除真实的物理空间。那么，remove 如何删除呢？前后结果如何发生变化？看一下图 9.1 就知道了。

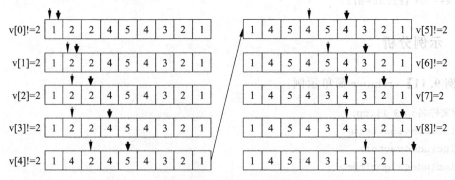

图 9.1　remove 过程图例

图9.1中两个指针同时指向向量v1头。(1)v[0]=1,v[0]!=2,则两个指针分别加1,指向下一个元素。(2)v[1]=2,则粗指针加1,指向下一个元素,细指针不变。(3)v[2]=2,则粗指针加1,指向下一个元素,细指针不变。(4)v[3]=4,v[3]!=2,则细指针位置元素由粗指针位置元素代替,即用4代替。之后,粗细指针均加1,指向一个元素。(5)v[4]=5,v[4]!=2,则细指针位置元素由粗指针位置元素代替,即用4代替。之后,粗细指针均加1,指向一个元素。同学们可继续往下分析,直到得出结果。

从图9.1可知,remove删除功能只实现元素的覆盖,细指针起始位置到结束位置间的元素才是真正的结果。对比示例:粗指针起始位置=细指针起始位置=v1.begin(),粗指针结束位置=v1.end(),细指针结束位置= remove(v1.begin(),v1.end(),2)。

【例9.12】 remove_copy简单示例。

```
//文件名:e9_12.cpp
#include<iostream>
#include<vector>
#include<algorithm>
#include<iterator>
using namespace std;

int main()
{
    int a[]={1,2,3,4,5};
    cout<<"原始数据a[]=";
    copy(a, a+5, ostream_iterator<int>(cout, "\t"));
    cout<<endl;

    vector<int>v1(a, a+5);                                              //修改容器自身
    vector<int>::iterator last1=remove_copy(v1.begin(),v1.end(),v1.begin(),3);
    cout<<"移去3后有效数据:";
    copy(v1.begin(), last1, ostream_iterator<int>(cout, "\t"));
    cout<<endl;

    vector<int>v2(a, a+5);
    vector<int>v3;
    cout<<"移去3复制到另一容器数据:";
    remove_copy(v2.begin(), v2.end(), back_inserter(v3),3);             //自身容器没修改
    copy(v3.begin(), v3.end(), ostream_iterator<int>(cout, "\t"));
                                                                        //结果复制到另一容器
    cout<<endl;
    return 0;
}
```

此示例说明remove_copy可修改自身容器结果,或者把结果输出到另一新容器中。

【例 9.13】 已知学生信息包含姓名、学号、语文及数学成绩,从学生集合中提取总成绩大于或等于150分的学生信息并按文本格式存入文件 stud.dat 中。

```cpp
//文件名:e9_13.cpp
#include <iostream>
#include <fstream>
#include <vector>
#include <string>
#include <algorithm>
#include <functional>
#include <iterator>
using namespace std;

class Student
{
public:
    string name;                    //姓名
    string studno;                  //学号
    int chinese;                    //语文
    int math;                       //数学
public:
    Student()
    {
    }
    Student(string name,string studno,int chinese, int math)
    {
        this->name=name, this->studno=studno;
        this->chinese=chinese, this->math=math;
    }
    bool operator<(const int & total) const
    {
        return ((chinese+math)<total);
    }
};
bool MyCompare(const Student & s)
{
    return s<150;                   //总成绩是否大于150
}
ostream & operator<<(ostream & os, const Student & s)
{
```

```
    os<<s.name<<"\t"<<s.chinese<<"\t"<<s.math;
    return os;
}

int main()
{
    Student s1("zhang","1001",60,70);
    Student s2("li", "1002",70,80);
    Student s3("zhao", "1003",75,85);
    Student s4("wang", "1004",68,78);
    Student s5("zhou", "1005",86,76);
    Student s6("qian", "1006",30,80);

    vector<Student>v;                    //生成学生向量集合
    v.push_back(s1), v.push_back(s2);
    v.push_back(s3), v.push_back(s4);
    v.push_back(s5), v.push_back(s6);

    ofstream out("d:\\stud.dat");        //创建文件
    remove_copy_if(v.begin(), v.end(), ostream_iterator<Student>(out, "\n"),
    MyCompare);
    out.close();
    return 0;
}
```

（1）本题主要是演示 remove_copy_if 功能，因此学生信息集合直接在 main 函数中前半部分生成了。根据题意，可能很多同学会想到先用非变异函数中的查询函数查询哪些学生总成绩大于 150，然后再把它存入文件，思路是正确的，但不可能用一条语句完成，而用 remove_copy_if 函数仅用一条语句就完成了查询及文件输出功能。因此该函数虽然划分为删除函数，但它也具有查询功能。所以如果按某种其他方式分类，也许可将 copy、replace_copy、remove_copy 划分成一类，find、remove_copy 划分成一类，这一点同学们在学习中要加以体会，要把知识学活。

（2）在全局函数 MyCompare 中函数体是 return s<150，大于号两侧是不同类型，左侧是 Student 对象，右侧是整数类型，因此要重载基本类 Student 中操作符 operator<。要查询的是成绩大于 150 分的学生，按正常情况来说，应该把上述小于号改成大于号。可以这样理解："查询结果是大于或等于 150 分的学生"相当于"移走小于 150 分的学生"，而这正好与 remove 系列函数的语义相符，因此本题中用到的都是小于号。

（3）如果想把查询结果放入另外一个向量，而不是文件，如下代码即可实现。

```
vector<Student>vResult;
remove_copy_if(v.begin(), v.end(), back_inserter(vResult),MyCompare);
```

9.8 唯一

9.8.1 主要函数

主要函数有：
- unique()：删除相邻的重复元素。
- unique_copy()：复制序列时删除相邻的重复元素。

函数原型如下所示。

① unique

原型：

```
template<class FwdIt>
    FwdIt unique(FwdIt first, FwdIt last);
template<class FwdIt, class Pred>
    FwdIt unique(FwdIt first, FwdIt last, Pred pr);
```

参数说明：
- FwdIt：前向迭代器。
- Pred：二元判定函数。

第一个模板函数先令 X=first，返回值是 X。

```
if (N==0||!(*(first+N)==V))    N∈[0,last-first)
    V=*(first+N), *X++=V;
```

第二个模板函数先令 X=first，返回值是 X。与前一模板函数功能相近，只不过调用了一元判定函数。

```
if (N==0||!(pr(*(first+N))==V))    N∈[0,last-first)
    V=*(first+N), *X++=V;
```

② unique_copy

原型：

```
template<class InIt, class OutIt>
    OutIt unique_copy(InIt first, InIt last, OutIt x);
template<class InIt, class OutIt, class Pred>
    OutIt unique_copy(InIt first, InIt last, OutIt x, Pred pr);
```

参数说明：
- InIt：输入迭代器。
- OutIt：输出迭代器。

- Pred：二元判定函数。

第一个模板函数：

```
if (N==0||!(*(first+N)==V))        N∈[0,last-first)
V=*(first+N), *x++=V;
```

第二个模板函数：

```
if (N==0||!pr(*(first+N),V))        N∈[0,last-first)
V=*(first+N), *x++=V;
```

9.8.2 示例分析

【例 9.14】 unique 函数简单示例。

```
//文件名：e9_14.cpp
#include<iostream>
#include<vector>
#include<algorithm>
#include<iterator>
using namespace std;
int main()
{
    int a[]={1,2,2,3,4,2,2,5};
    vector<int>v1(a, a+8);
    cout<<"原始向量 v1:";
    copy(v1.begin(), v1.end(), ostream_iterator<int>(cout, "\t"));
    cout<<endl;

    vector<int>::iterator last=unique(v1.begin(), v1.end());

    cout<<"unique 后向量 v1:";
    copy(v1.begin(), v1.end(), ostream_iterator<int>(cout, "\t"));
    cout<<endl;
    cout<<"unique 后向量 v1 有效元素:";
    copy(v1.begin(), last, ostream_iterator<int>(cout, "\t"));
    cout<<endl;
    return 0;
}
```

执行结果是：

```
原始向量 v1:            1 2 2 3 4 2 2 5
unique 后向量 v1:       1 2 3 4 2 5 2 5
```

unique 后向量 v1 有效元素：1 2 3 4 2 5

（1）unique 函数功能是消除容器的相邻相等元素，若有连续 N 个重复元素，则只保留第 1 个出现的元素。通过与结果比较，可知容器 v1 的大小没有发生变化，并不是真正的消除，只是相邻的多余重复元素被后续元素覆盖了。要遍历 unique 后的有效元素，一定要利用 unique 函数执行结束后的返回值 vector<int>::iterator last=unique(v1.begin()，v1.end())，则[v1.begin()，last)间指向元素才是 unique 后的有效元素，而[last，v1.end())是非有效元素。

（2）原始序列是"1 2 2 3 4 2 2 5"，2 是重复元素，一般来说，希望消除多余的 2 后应该得到序列"1 2 3 4 5"，这一结果与示例中结果不同，因此 unique 函数一般应用在容器元素排好序的情况下。如原始向量经排序后为"1 2 2 2 2 3 4 5"，再执行 unique 函数，结果即是"1 2 3 4 5"。只是现在还没有学到 STL 排序函数。

【例 9.15】 文本文件存放单词（格式行如表 9.2 所示），已排好序，请统计有多少个单词（忽略大小写）。

表 9.2 文本文件格式示例

| How how how how |
| how how how |
| Is is |
| You you you you |

```
//文件名：e9_15.cpp
#include<iostream>
#include<fstream>
#include<vector>
#include<algorithm>
#include<string>
#include<iterator>
using namespace std;

bool MyStrCompare(string s1, string s2)
{
    bool bRet=false;
    int value=stricmp(s1.c_str(), s2.c_str());
    if(value==0)
        bRet=true;

    return bRet;
}

int main()
{
    ifstream out("d:\\data.txt");
    vector<string>v;                          //读文本文件，按字符串读
    copy(istream_iterator<string>(out), istream_iterator<string>(),back_inserter(v));
    cout<<endl;

    cout<<"文本文件中单词有：";
    copy(v.begin(), v.end(), ostream_iterator<string>(cout, "\t"));
    cout<<endl;

    vector<string>vstr;
```

```
        cout<<"文本文件中去掉重复单词后:";
        unique_copy(v.begin(), v.end(), back_inserter(vstr), MyStrCompare);
        copy(vstr.begin(), vstr.end(), ostream_iterator<string>(cout, "\t"));
        cout<<endl;
        return 0;
}
```

自定义函数 MyStrCompare 中可以人为定义两字符串是否相等,更加灵活。本例中运用 stricmp 系统函数实现了忽略大小写比较字符串是否相等的功能。

9.9 反转

9.9.1 主要函数

主要函数有:
- reverse():反转元素的次序。
- reverse_copy():复制序列时反转元素的次序。

函数原型如下所示。

① reverse

原型:

```
template<class BidIt>
void reverse(BidIt first, BidIt last);
```

参数说明:

BidIt:双向迭代器。

该函数功能是:在内部循环调用 swap 函数。

swap(*(first+N), *(last-1-N))   N∈[0,(last-first)/2)

② reverse_copy

原型:

```
template<class BidIt, class OutIt>
OutIt reverse_copy(BidIt first, BidIt last, OutIt x);
```

参数说明:

- BidIt:双向迭代器。
- OutIt:输出迭代器。

该函数功能是:

(x+N)=(last-1-N)   N∈[0,last-first)

返回 x+(last-first)。

9.9.2 示例分析

【例 9.16】 反转函数简单示例。

```cpp
//文件名:e9_16.cpp
#include<iostream>
#include<vector>
#include<algorithm>
#include<iterator>
using namespace std;
int main()
{
    int a[]={1,2,3,4,5};

    vector<int>v1(a, a+5);
    cout<<"原始数据向量 v1:";
    copy(v1.begin(), v1.end(), ostream_iterator<int>(cout, "\t"));
    cout<<endl;

    cout<<"反转向量 v1(reverse):";
    reverse(v1.begin(), v1.end());
    copy(v1.begin(), v1.end(), ostream_iterator<int>(cout, "\t"));
    cout<<endl;

    vector<int>v2(a, a+5);
    reverse_copy(v2.begin(), v2.end(), v2.begin());//--------------------- (a)
    cout<<"反转向量 v2-->v2(reverse_copy):";
    copy(v2.begin(), v2.end(), ostream_iterator<int>(cout, "\t"));
    cout<<endl;

    vector<int>v3(a, a+5);
    vector<int>v4;
    reverse_copy(v3.begin(), v3.end(), back_inserter(v4));
    cout<<"反转向量 v3-->v4(reverse_copy):";
    copy(v4.begin(), v4.end(), ostream_iterator<int>(cout, "\t"));
    cout<<endl;
    return 0;
}
```

执行结果是:

原始数据向量 v1:   1 2 3 4 5
反转向量 v1(reverse):   5 4 3 2 1

反转向量 v2-->v2(reverse_copy)：5 4 3 4 5
反转向量 v3-->v4(reverse_copy)：5 4 3 2 1

原始数据是"1 2 3 4 5"，翻转后应该是"5 4 3 2 1"。对比结果有两种情况是正确的：(1)直接利用 reverse 函数；(2)利用 reverse_copy 函数，原始容器是 v3，翻转结果存储容器是 v4，v3 和 v4 是不同的容器。错误的结果代码见(a)，特点是 reverse_copy 函数的原始容器与翻转结果存储容器是同一个 v2。

其实，看一下函数原型就会发现：reverse 与 reverse_copy 算法是不同的，reverse 函数主要是交换算法 swap 的循环，reverse_copy 根本没有 swap 过程，仅是赋值。对本题而言，有 5 个元素，执行 reverse，swap 交换 2 次；执行 reverse_copy，赋值 5 次。因此，在某些情况下结果一定不同。这和以前学过的算法有些不同，如 unique、unique_copy，它们在本质上是相同的，所以假设对某容器 V 而言：unique(V.begin(), V.end())与 unique(V.begin(), V.end(), V.begin(), MyCompare)结果是一样的(MyCompare 是二元函数)。

9.10 环移

9.10.1 主要函数

主要函数有：
- rotate()：循环移动元素。
- rotate_copy()：复制序列时循环移动元素。

函数原型如下所示。
① rotate
原型：

template<class FwdIt>
void **rotate**(FwdIt first, FwdIt middle, FwdIt last);

参数说明：
FwdIt：前向迭代器。
该函数功能是：一方面把 *[middle, last)间元素依次存储到 *[first, first+last−middle)中，另一方面把原始 *[first, middle)间元素依次存储到 *[last−middle, last)中。
② rotate_copy
原型：

template<class FwdIt, class OutIt>
OutIt **rotate_copy**(FwdIt first, FwdIt middle, FwdIt last, OutIt x);

参数说明：
- FwdIt：前向迭代器。

- OutIt：输出迭代器。

该函数功能是：一方面把 * [middle, last) 间元素依次存储到 * [x, x+(last-middle)) 中,另一方面把 * [first, middle) 间元素依次存储到 * [x+(last-middle), x+(last-first)) 中。

9.10.2 示例分析

【例 9.17】 已知序列 1,2,3,4,5,6,7,8。要求每间隔 1s 循环显示一次,当按下任意键时,退出演示。

```
//文件名: e9_17.cpp
#include<iostream>
#include<vector>
#include<algorithm>
#include<conio.h>
#include<time.h>
#include<iterator>
using namespace std ;

void delay()                                        //1s 延时程序
{
    int start=time(NULL);
    int end=start;
    do
    {
        if(start !=end)                             //当 start!=end 时时间过了 1s
            break;
    }while(end=time(NULL));
}

int main()
{
    int a[]={1,2,3,4,5,6,7,8};
    vector<int>v(a, a+8);
    cout<<endl;
    while(!kbhit())                                 //按任意键退出演示
    {
        copy(v.begin(), v.end(), ostream_iterator<int>(cout, "\t"));
        rotate(v.begin(), v.begin()+1, v.end());
        delay();
        cout<<endl;
    }
    return 0;
}
```

(1) 利用 rotate 函数很好地实现了循环显示,middle 位置选为 v.begin()+1。当第 1 次执行 rotate 时,middle 指向元素 2,所以结果是 2,3,4,5,6,7,8,1;当第 2 次执行 rotate 时,middle 指向 3,所以结果是 3,4,5,6,7,8,1,2。以下分析,以此类推。

(2) 延时程序 delay 应用了系统 time 函数,算法是:先获取当前时间秒数 start,接着在 while 循环中不断读结束秒数 end,当 start!= end 时一定是过了 1s。

(3) 按任意键退出用 kbhit 函数,用 getch 函数是不行的。

【例 9.18】 list 容器中环移函数用法。

```
//文件名:e9_18.cpp
int main()
{
    int a[]={1,2,3,4,5,6,7,8};
    cout<<"原始数据 a[]=";
    copy(a, a+8, ostream_iterator<int>(cout, "\t"));
    cout<<endl;

    list<int>v1(a, a+8);
    list<int>::iterator middle=v1.begin();
    advance(middle, 3);
    rotate(v1.begin(), middle, v1.end());
    cout<<"以 4 为中心环移(rotate):";
    copy(v1.begin(), v1.end(), ostream_iterator<int>(cout,"\t"));
    cout<<endl;

    list<int>v2(a, a+8);
    list<int>v3;
    middle=v2.begin();
    cout <<"以 4 为中心环移 v2-->v3(rotate_copy):";
    advance(middle, 3);
    rotate_copy(v2.begin(), middle, v2.end(), back_inserter(v3));
    copy(v3.begin(), v3.end(), ostream_iterator<int>(cout, "\t"));
    cout<<endl;
    return 0;
}
```

(1) 该例以序列{1,2,3,4,5,6,7,8}中 4 为中心,在 list 容器中进行环移操作。注意 rotate_copy 函数,本例中是 rotate_copy(v2.begin(), middle, v2.end(), back_inserter(v3)),表示把环移结果写入另外一个容器 v3 中。如果把 back_inserter(v3)改为 v2.begin(),那么相当于把环移结果写入自身容器 v2 中,这时与 rotate(v2.begin(),middle,v2.end())是等同的。

(2) 由于是 list 容器,它的内部迭代器是双向迭代器,元素 4 的迭代器位置不能用 middle=v2.begin()+3 表示,只有随机迭代器如 vector 中的迭代器才能执行 operate+(n) 操作,但是可以用系统函数 advance 轻松实现,这一点同学们要加强理解。

9.11 随机

9.11.1 主要函数

主要函数有：
random_shuffle()：采用均匀分布来随机移动元素。
函数原型如下所示。

```
template<class RanIt>
    void random_shuffle(RanIt first, RanIt last);
template<class RanIt, class Fun>
    void random_shuffle(RanIt first, RanIt last, Fun& f);
```

参数说明：
- RanIt：随机迭代器。
- Fun：一元函数对象。

第一个模板函数功能是：采用系统默认的随机数发生器打乱 *[first, last)间元素顺序，随机生成一个新排列。具体是 swap (*(first+N)，*(first+M))，N∈[1, last−first)，M 是系统产生的一个随机数，且 M∈[0, N)。

第二个模板函数功能是：采用自定义随机数发生器打乱 *[first, last)间元素顺序，随机生成一个新排列。具体是 swap (*(first+N)，*(first+M))，N∈[1, last−first)，M=f(N)。

传统的随机数产生方法是使用 ANSI C 的函数 rand()，然后格式化结果以便使结果落在指定范围内。但是，使用这个方法至少有两个缺点。首先，做格式化时，结果常常是扭曲的，所以得不到正确的随机数(如某些数的出现频率要高于其他数)。其次，rand()只支持整型数，不能用它来产生随机字符、浮点数、字符串或数据库中的记录。

对于以上的两个问题，random_shuffle()算法给出了更好的解决方案，用这种算法可以产生不同类型的随机数。产生指定范围内的随机元素集的最佳方法是创建一个顺序序列(也就是向量或者内置数组)，在这个顺序序列中含有指定范围的所有值，之后用 random_shuffle()算法打乱元素排列顺序。

9.11.2 示例分析

【例 9.19】 随机函数简单示例。

```
//文件名：e9_19.cpp
#include <iostream>
```

```cpp
#include <algorithm>
#include <vector>
#include <iterator>

using namespace std;
int main()
{
    int a[]={1,2,3,4,5};
    float b[]={1.1f,2.1f,3.1f,4.1f,5.1f};
    char c[]={'a','b','c','d','e'};

    cout<<"原始 a[]=";
    copy(a, a+5, ostream_iterator<int>(cout, "\t"));
    random_shuffle(a, a+5);
    cout<<"random_shuffle 后:";
    copy(a, a+5, ostream_iterator<int>(cout, "\t"));
    cout<<endl;

    cout<<"原始 b[]=";
    copy(b, b+5, ostream_iterator<float>(cout, "\t"));
    random_shuffle(b, b+5);
    cout<<"random_shuffle 后:";
    copy(b, b+5, ostream_iterator<float>(cout, "\t"));
    cout<<endl;

    cout<<"原始 c[]=";
    copy(c, c+5, ostream_iterator<char>(cout, "\t"));
    random_shuffle(c, c+5);
    cout<<"random_shuffle 后:";
    copy(c, c+5, ostream_iterator<char>(cout, "\t"));
    cout<<endl;
    return 0;
}
```

可以看出：利用 random_shuffle 实现了整型、浮点、字符序列的随机化，当然也可以是其他类型，前提是初始化序列必须已知。但是也有一个缺点：每次执行时产生的随机数序列是固定的，若想不固定，则要用到 random_shuffle 的第二个模板函数，见下例。

【例 9.20】 自定义随机数生成器。

//文件名：e9_20.cpp
#include <iostream>
#include <algorithm>
#include <vector>

```cpp
#include <conio.h>
#include <time.h>
using namespace std;
class MyRand
{
public:
    int operator()(int n)
    {
        srand(time(NULL));
        return rand()%n ;
    }
};

int main()
{
    int a[]={10,21,32,43,54,65,76,87,98};
    vector<int>v(a, a+5);
    MyRand obj;
    random_shuffle(v.begin(), v.end(), obj);
    copy(v.begin(), v.end(), ostream_iterator<int>(cout, "\t"));
    cout<<endl;
    return 0;
}
```

程序每次执行时向量 v 元素序列都不同,关键是在一元函数 MyRand 类中,利用 srand(time(NULL))使每次产生的随机数种子都不同。另外与上文讲的函数原型对比,"int operator()(int n)"中参数 n 不是向量 v 中的序列元素,而是[1,v.size()),这一点尤其重要。

因此可以根据需要写对应的一元随机函数类,例如若容器有 total 个元素,要求[0,total/2)可以随机变化,[total/2,total)可以随机变化,但不可以越界。代码如下:

```cpp
class MyRand
{
    int total;
public:
    MyRand(int total)
    {
        this->total=total;
    }
    int operator()(int n)
    {
        srand(time(NULL));
        int m=rand()%n;
        if(n>=total/2 && m<total/2)     //如果 n 是后半部分元素索引,而 m 随机数是
                                        [0,total/2)
```

```
            {
                m+=total/2;
                //则 m+=total/2,保证 m 是后半部分元素索引才能保证是后半部分元素间互相交换
            }
            return m;
        }
};
int main()
{
    int a[]={1,2,3,4,5,6,7,8,9};
    vector<int>v(a, a+9);

    MyRand obj(9);
    char ch;
    while((ch=getch())!='a')
    {
        random_shuffle(v.begin(), v.end(), obj);
        copy(v.begin(), v.end(), ostream_iterator<int>(cout, "\t"));
        cout<<endl;
    }
    return 0;
}
```

看执行结果就会发现,向量 v 的 9 个元素序列中,前 4 个元素是{1,2,3,4}的随机排列,后 5 个元素是{5,6,7,8,9}的随机排列。

9.12 划分

9.12.1 主要函数

主要函数有:
- partition():将满足某谓词的元素都放到前面。
- stable_partition():将满足某谓词的元素都放到前面并维持原顺序。

函数原型如下所示。

① partition

原型:

```
template<class BidIt, class Pred>
BidIt partition(BidIt first, BidIt last, Pred pr);
```

参数说明：
- BidIt：双向迭代器。
- Pred：一元判定函数。

该函数功能是：容器元素重新定位，确定位置 K。当 0＜N＜K 时，pr(* (first＋N))是 true；当 K≤N＜last－first 时，pr(* (first＋N))是 false。函数最终返回 first＋K。

② stable_partition

原型：

template<class FwdIt, class Pred>
FwdIt **stable_partition**(FwdIt first, FwdIt last, Pred pr);

参数说明：
- BidIt：双向迭代器。
- Pred：一元判定函数。

该函数功能是：与 partition 功能相似，只不过是稳定划分。当容器中两个元素同时满足条件或同时不满足条件时，原先在前面的元素当重新定位后仍在前面；如果一个元素满足条件另一个元素不满足条件时，则重新定位后元素的位置前后顺序不能确定。

9.12.2 示例分析

【例 9.21】 划分函数简单示例。

```
//文件名：e9_21.cpp
#include<iostream>
#include<algorithm>
#include<vector>
#include<functional>
#include<iterator>
using namespace std;
int main()
{
    int a[]={1,7,2,5,3,4,8,2,3,6};
    vector<int>v1(a, a+10);
    cout<<"原始数据：";
    copy(v1.begin(), v1.end(), ostream_iterator<int>(cout, "\t"));
    cout<<endl;

    cout<<"按<4条件划分(partition):";                    //<4元素放在容器前面
    partition(v1.begin(),v1.end(), bind2nd(less<int>(), 4));
    copy(v1.begin(), v1.end(), ostream_iterator<int>(cout, "\t"));
    cout<<endl;
```

```
        cout<<"按<4条件划分(stable_partition):";
        stable_partition(v1.begin(),v1.end(),bind2nd(less<int>(),4));
                                                                    //<4元素放在容器前面
        copy(v1.begin(), v1.end(), ostream_iterator<int>(cout, "\t"));
        cout<<endl;
        return 0;
    }
```

执行结果是：

原始数据：   1 7 2 5 3 4 8 2 3 6
按<4条件划分(partition)：   1 3 2 2 3 4 8 5 7 6
按<4条件划分(stable_partition):1 3 2 2 3 4 8 5 7 6

从结果看，partition、stable_partition 操作结果是一样的，那这两个函数有什么区别呢？请看下例。

【例 9.22】 已知学生信息包含姓名和成绩两项，把学生集合中成绩小于 75 分的放在容器的前面。

```
//文件名：e9_22.cpp
#include <iostream>
#include <algorithm>
#include <vector>
#include <string>
#include <algorithm>
#include <functional>
#include <iterator>
using namespace std;
struct Student
{
    char name[20];
    int grade;
};
bool GradeCmp(Student& s)
{
    return s.grade < 75;
}
ostream& operator<< (ostream& os, const Student& s)
{
    os<<"("<<s.name<<"\t"<<s.grade<<")";
    return os;
}
int main()
{
    Student s[]={{"li1",79},{"li2",70},{"li3",68},{"li4",78},{"li5",65}};
    cout<<"学生原始序列:"<<endl;
```

```cpp
        copy(s, s+5, ostream_iterator<Student>(cout, "\t"));
        cout<<endl;

        vector<Student>v1(s, s+5);
        cout<<"不稳定划分,成绩<75(partition):"<<endl;
        partition(v1.begin(), v1.end(), GradeCmp);
        copy(v1.begin(), v1.end(), ostream_iterator<Student>(cout, "\t"));
        cout<<endl;

        vector<Student>v2(s, s+5);
        cout<<"稳定划分,成绩<75(stable_partition):"<<endl;
        stable_partition(v2.begin(), v2.end(), GradeCmp);
        copy(v2.begin(), v2.end(), ostream_iterator<Student>(cout, "\t"));
        cout<<endl;
        return 0;
    }
```

执行结果是:

```
学生原始序列:                          (li1 79) (li2 70) (li3 68) (li4 78) (li5 65)
不稳定划分,成绩<75(partition):         (li5 65) (li2 70) (li3 68) (li4 78) (li1 75)
稳定划分,成绩<75(stable_partition): (li2 70) (li3 68) (li5 65) (li1 79) (li4 78)
```

从结果可清晰看出 partition 是不稳定划分,stable_partition 是稳定划分。

【例 9.23】 list 整型容器按小于 4 条件划分成两部分,并求有多少元素小于 4。

```cpp
//文件名:e9_23.cpp
#include <iostream>
#include <list>
#include <algorithm>
#include <functional>
using namespace std;
int main()
{
    int a[]={1,7,3,6,4,10,9,5,2,8};
    list<int>v(a, a+10);
    list<int>::iterator mid=partition(v.begin(), v.end(), bind2nd(less<int>(), 4));
    int nSize=distance(v.begin(), mid);
    cout<<nSize<<endl;
    return 0;
}
```

先利用 partition 函数返回值得到划分后符合小于 4 条件的尾迭代器指针 mid,再巧妙利用 distance(v.begin(), mid) 即得到有多少元素小于 4。

第 10 章 排序及相关操作

主要操作函数如表 10.1 所示。

表 10.1 排序及相关函数列表

序号	功能	函数名称	说明
1	排序	sort	以很好的平均效率排序
		stable_sort	稳定排序，维持相同元素的原有顺序
		partial_sort	局部排序
		partial_sort_copy	复制的同时进行局部排序
2	第 n 个元素	nth_element	将第 n 个元素放到它的正确位置
3	二分检索	lower_bound	找到大于等于某值第一次出现的迭代器位置
		upper_bound	找到大于某值第一次出现的迭代器位置
		equal_range	找到等于某值迭代器范围
		binary_search	在有序序列中确定给定元素是否存在
4	归并	merge	归并双迭代器两个有序序列
		inplace_merge	归并单迭代器两个接续的有序序列
5	序结构上的集合操作	includes	一个序列为另一个序列的子序列时为真
		set_union	构造两个集合的有序并集
		set_intersection	构造两个集合的有序交集
		set_difference	构造两个集合的有序差集
		set_symmetric_difference	构造两个集合的有序对称差集（并—交）
6	堆操作	make_heap	从序列构造堆
		pop_heap	从堆中弹出元素
		push_heap	向堆中加入元素
		sort_heap	给堆排序

续表

序号	功能	函数名称	说明
7	最大和最小	min	两个值中较小的
		max	两个值中较大的
		min_element	序列中的最小元素
		max_element	序列中的最大元素
8	词典比较	lexicographical_compare	两个序列按字典序比较
9	排列生成器	next_permutation	按字典序的下一个排列
		prev_permutation	按字典序的上一个排列
10	数值算法	accumulate	累积求和
		inner_product	内积求和
		partial_sum	创建新序列,每个元素值代表指定范围内该位置前所有元素之和
		adjacent_difference	获得相邻元素差集

10.1 排序

10.1.1 主要函数

- sort():以很好的平均效率排序。
- stable_sort():稳定排序,维持相同元素的原有顺序。
- partial_sort():局部排序。
- partial_sort_copy():复制的同时进行局部排序。

函数原型如下所示。

① sort

原型:

```
template<class RanIt>
    void sort(RanIt first, RanIt last);
template<class RanIt, class Pred>
    void sort(RanIt first, RanIt last, Pred pr);
```

参数说明:

- RanIt:随机迭代器,first 表示起始迭代器指针,last 表示终止迭代器指针。
- Pred:普通全局函数或二元函数。

第一个模板函数[first,last)间迭代器指示的元素数据默认按升序排列;第二个模板函数定义了比较函数 pr(x,y)代替 operator<(x,y)，功能是相似的，属于不稳定排序。

② stable_sort

原型：

```
template<class RanIt>
    void stable_sort(RanIt first, RanIt last);
template<class RanIt, class Pred>
    void stable_sort(RanIt first, RanIt last, Pred pr);
```

参数说明：

- RanIt：随机迭代器，first 表示起始迭代器指针，last 表示终止迭代器指针。
- Pred：普通全局函数或二元函数。

第一个 sort 函数[first,last)间迭代器指示的元素数据默认按升序排列，第二个 sort 函数定义了比较函数 pr(x,y)代替 operator<(x,y)，功能是相似的。与 sort 函数相比，和它的名字一样，属于稳定排序。

③ partial_sort

原型：

```
template<class RanIt>
    void partial_sort(RanIt first, RanIt middle, RanIt last);
template<class RanIt, class Pred>
    void partial_sort(RanIt first, RanIt middle, RanIt last, Pred pr);
```

参数说明：

- RanIt：随机迭代器，first 表示起始迭代器指针，last 表示终止迭代器指针，middle 表示[first,last)间迭代器。
- Pred：普通全局函数或二元函数。

该函数实现了局部元素排序功能。对[first，last)间的元素排序结束后，仅前 middle−first−1 个元素是必须按要求排好序的，其他元素不一定是排好序的。即对任意 N∈[0，middle−first]，M∈(N,last−first)，都有 *(first+N)< *(first+M)。第二个函数与第一个函数相比定义了比较函数 pr(x,y)代替 operator<，功能是相似的。

④ partial_sort_copy

原型：

```
template<class InIt, class RanIt>
    RanIt partial_sort_copy(InIt first1, InIt last1,RanIt first2, RanIt last2);
template<class InIt, class RanIt, class Pred>
    RanIt partial_sort_copy(InIt first1, InIt last1,RanIt first2, RanIt last2, Pred pr);
```

参数说明：

- InIt：输入迭代器。

- RanIt：随机迭代器。
- Pred：普通全局函数或二元函数。

该函数功能是：与 partial_sort 相比主要有两点不同：（1）排序结果可以输出到另外一个容器（当然也可以是自身容器）；（2）partial_sort 函数中直接给出了 middle 值，而该函数 middle 值是计算出来的，middle ＝ min(last1－first1, last2－first2)＋first1。之后，对任意 N∈[0, middle－first1]，M∈(N, last1－first1)，都有 *(first2＋N)＜*(first1＋M)。第二个函数与第一个函数相比定义了比较函数 pr(x, y) 代替 operator＜，功能是相似的。

10.1.2 示例分析

【例 10.1】利用 sort 对整型向量升序排序。

```cpp
//文件名：e10_1.cpp
#include <iostream>
#include <algorithm>
#include <vector>
#include <iterator>
using namespace std;
int main(int argc, char * argv[])
{
    int a[]={1,8,6,10,4};
    vector<int>v(a, a+5);
    sort(v.begin(), v.end());
    cout<<"升序排序结果是："<<endl;
    copy(v.begin(), v.end(), ostream_iterator<int>(cout, "\t"));    //1 4 6 8 10
    return 0;
}
```

可以看出，本例用第一个 sort 模板函数完成了升序排列。如果要求降序排列就不行了，就要用到第二个 sort 模板函数。

```cpp
bool myless(int &m, int &n)
{
    return m>n;
}
int main(int argc, char * argv[])
{
    int a[]={1,8,6,10,4};
    vector<int>v(a, a+5);
    sort(v.begin(), v.end(), myless);
    cout <<"降序排序结果是：" <<endl;
    copy(v.begin(), v.end(), ostream_iterator<int>(cout, "\t"));
```

```
        return 0;
    }
```

通过自定义全局函数 myless 完成了向量的降序排序，当然可以更简化，直接用系统 greater 二元函数就可以了，即 sort(v.begin(), v.end(), greater<int>())。

以上都是对基本数据类型而言的，若对学生成绩进行排序，又该如何呢？程序见下例。

【例 10.2】 对学生成绩进行升序排列。

```
//文件名：e10_2.cpp
#include <iostream>
#include <algorithm>
#include <vector>
#include <string>
#include <iterator>
using namespace std;
class Student
{
public:
    int NO;                        //学号
    string name;
    int grade;                     //成绩
    Student(int NO, string name, int grade)
    {
        this->NO=NO;
        this->name=name;
        this->grade=grade;
    }
    bool operator< (const Student &s) const
    {
        return grade <s.grade;
    }
};
ostream & operator<< (ostream & os, const Student & s)
{
    os<<s.NO<<"\t"<<s.name<<"\t"<<s.grade;
    return os;
}
int main(int argc, char * argv[])
{
    Student s1(101,"张三", 90);
    Student s2(102,"李四", 80);
    Student s3(103, "王五", 85);
    Student s4(103, "赵六", 65);
```

```
    vector<Student>v;
    v.push_back(s1);
    v.push_back(s2);
    v.push_back(s3);
    v.push_back(s4);
    sort(v.begin(),v.end());
    cout<<"升序排序结果是："<<endl;
    cout<<"学号\t"<<"姓名\t"<<"成绩"<<endl;
    copy(v.begin(),v.end(),ostream_iterator<Student>(cout,"\n"));
    return 0;
}
```

可以看出，必须重载 Student 类中的"<"运算符，而且参数只能是 Student 对象的引用，如何进行排序主要是在该函数中完成的。在 main 函数中，sort 函数是采用第一个模板函数完成排序功能的。若用第二个 sort 模板函数，对此例应做如下修改：(1)主函数中 sort 函数改为 sort(v.begin()，v.end()，myascend)；(2)去掉 Student 类中的 bool operator<(Student & s)函数，在外部增加一个普通的全局函数，例如：

```
bool myascend(Student & s1, Student & s2)
{
    return s1.grade<s2.grade;
}
```

或者直接用 less 函数对象即可，基本类 Student 不用做任何修改，仅主程序 sort 函数改为下述就可以了。

```
sort(v.begin(),v.end(),less<Student>())
```

【例 10.3】 利用 partial_sort 取整型向量最小的 4 个元素。

```
//文件名：e10_3.cpp
#include<iostream>
#include<algorithm>
#include<vector>
#include<iterator>
using namespace std;
int main()
{
    int a[]={10,1,3,9,7,6,2,4,5,8};
    vector<int>v(a, a+10);
    cout<<"原始向量数据：";
    copy(v.begin(),v.end(),ostream_iterator<int>(cout,"\t"));
    cout<<endl;
    partial_sort(v.begin(),v.begin()+4,v.end());
    cout<<"partial_sort 后(前 4 个元素按升序排列)：";
    copy(v.begin(),v.end(),ostream_iterator<int>(cout,"\t"));
```

```
        cout<<endl;
        return 0;
}
```

执行结果是：

原始向量数据：10 1 3 9 7 6 2 4 5 8
partial_sort 后(前 4 个元素按升序排列)：1 2 3 4 10 9 7 6 5 8

可以看出前 4 个元素是排好序的，且是向量中最小的 4 个元素，而其他元素未必是排好序的。

【例 10.4】 求成绩最好的 3 位同学。

```
//文件名：e10_4.cpp
#include <iostream>
#include <vector>
#include <string>
#include <functional>
#include <algorithm>
using namespace std;
class Student
{
public:
    int NO;                //学号
    string name;
    int grade;             //成绩
    Student() {}
    Student(int NO, string name, int grade)
    {
        this->NO=NO;
        this->name=name;
        this->grade=grade;
    }
    bool operator>(const Student &s) const
    {
        return grade >s.grade;
    }
};
int main()
{
    vector<Student>v ;
    Student s1(101, "aaa", 90);
    Student s2(102, "bbb", 80);
    Student s3(103, "ccc", 60);
    Student s4(104, "ddd", 96);
    Student s5(105, "eee", 95);
    Student s6(106, "fff", 78);
```

```
        Student s7(107, "ggg", 94);
        Student s8(108, "hhh", 92);

        v.push_back(s1); v.push_back(s2);
        v.push_back(s3); v.push_back(s4);
        v.push_back(s5); v.push_back(s6);
        v.push_back(s7); v.push_back(s8);

        vector<Student>vResult(3);              //保存3个最好成绩学生的向量
        partial_sort_copy(v.begin(), v.end(), vResult.begin(), vResult.end(),greater
        <Student>());
        for(int i=0; i<vResult.size(); i++)
        {
            Student & s=vResult.at(i);
            cout<<s.NO<<"\t"<<s.name<<"\t"<<s.grade <<endl;
        }
        return 0;
    }
```

(1) Student 是一个类，要对它的各个对象降序排序，应该重载 operator> 运算符，本例中其函数体内容仅一行 return grade>s.grade，完成的是求最好的 3 个学生的成绩。若求成绩最差的 3 个学生，则应重载 operator< 运算符，函数体改为 return grade<s.grade，当然 partial_sort_copy 最后一个参数改为 less<Student>() 就可以了。

(2) 本例采用 partial_sort_copy，把排好序的有效数据复制到另外一个向量容器 vResult 中，用到了函数对象 greater<Student>()。若用自定义函数，仅做下述改动就可以了：主函数中 partial_sort_copy 改为 partial_sort_copy(v.begin(), v.begin()+3, v.end(), topgrade);，然后去掉 Student 类中的 operator> 运算符函数，增加一个全局函数 topgrade，如下所示：

```
    bool topgrade(Student& s, Student& t)
    {
        return s.grade>t.grade;
    }
```

【例 10.5】 list 容器排序问题。

```
//文件名：e10_5.cpp
#include <iostream>
#include <list>
#include <algorithm>
#include <iterator>
using namespace std;
int main()
{
```

```
    int a[]={10,1,3,9,7,6,2,4,5,8};
    list<int>l(a, a+10);
    //sort(l.begin(), l.end());                                  //这一行是错误的
    l.sort();                                                    //这一行正确
    copy(l.begin(), l.end(), ostream_iterator<int>(cout, "\t")); //1 2 3 4 5 6 7 8 9 10
    return 0;
}
```

(1) 注释行的程序是错误的,说明 list 容器不能用 sort 通用排序算法。这是由于 sort 需要的是随机迭代器,方便排序算法中的数据交换,而 list 提供的仅是双向迭代器。因此要想排序,只能用 list 类本身提供的 sort 函数,它有两种形式,已经在 7.4 节简单讲过。如果想让示例中元素降序排列,如下调用就可以了:l.sort(greater<int>())。当然也可以自定义二元函数。

(2) 假设先已知 list 容器,如何对它进行部分排序呢?直接应用 partial_sort 肯定是不行的。可以先把 list 容器元素复制到 vector 容器中,对 vector 容器进行 partial_sort,从而得到 list 容器部分排序的结果。

10.2 第 n 个元素

10.2.1 主要函数

nth_element():将第 n 个元素放到它的正确位置。

函数原型如下所示。

```
template<class RanIt>
    void nth_element(RanIt first, RanIt nth, RanIt last);
template<class RanIt, class Pred>
    void nth_element(RanIt first, RanIt nth, RanIt last, Pred pr);
```

参数说明:

- RanIt:随机迭代器,first 表示起始迭代器指针,last 表示终止迭代器指针。
- Pred:普通全局函数或二元函数。

该函数的功能是:在[first, last]指示的元素中,找第 n 个满足条件的元素,结果反映在 RanIt nth 表示的迭代器指针中。例如,班上有 10 个学生,想知道分数排在倒数第 4 名的学生。如果要满足上述需求,可以用 sort 排好序,然后取第 4 位(因为是由小到大排)。更聪明的会用 partial_sort,只排前 4 位,然后得到第 4 位。其实这时还是浪费时间,因为前两位根本没有必要排序,此时需要 nth_element。两个函数功能相近,第一个默认重载 operate<,第二个可定义二元函数对象。

10.2.2 示例分析

【例 10.6】 求第 3 个成绩最好的学生。

```cpp
//文件名：e10_6.cpp
#include<iostream>
#include<algorithm>
#include<vector>
#include<string>
#include<functional>
using namespace std;
class Student
{
public:
    int NO;              //学号
    string name;
    int grade;           //成绩
    Student(int NO, string name, int grade)
    {
        this->NO=NO;
        this->name=name;
        this->grade=grade;
    }
    bool operator>(const Student &s) const
    {
        return grade >s.grade;
    }
};
int main()
{
    vector<Student>v;
    Student s1(101, "aaa", 90);
    Student s2(102, "bbb", 80);
    Student s3(103, "ccc", 60);
    Student s4(104, "ddd", 96);
    Student s5(105, "eee", 95);
    Student s6(106, "fff", 78);
    Student s7(107, "ggg", 94);
    Student s8(108, "hhh", 92);
    v.push_back(s1); v.push_back(s2);
    v.push_back(s3); v.push_back(s4);
    v.push_back(s5); v.push_back(s6);
    v.push_back(s7); v.push_back(s8);
```

```
nth_element(v.begin(), v.begin()+2, v.end(),greater<Student>());
vector<Student>::iterator it=v.begin()+2;
cout << "第 3 个成绩好的学生信息为:" <<endl;
cout<<"学号\t姓名\t成绩"<<endl;
cout<<(*it).NO<<"\t"<<(*it).name<<"\t"<<(*it).grade<<endl;
return 0;
}
```

10.3 二分检索

10.3.1 主要函数

- lower_bound()：找到大于等于某值第一次出现的迭代器位置。
- upper_bound()：找到大于某值第一次出现的迭代器位置。
- equal_range()：找到等于某值迭代器范围。
- binary_search()：在有序序列中确定给定元素是否存在。

函数原型如下所示。

① lower_bound

原型：

```
template<class FwdIt, class T>
    FwdIt lower_bound(FwdIt first, FwdIt last, const T& val);
template<class FwdIt, class T, class Pred>
    FwdIt lower_bound(FwdIt first, FwdIt last, const T& val, Pred pr);
```

参数说明：

- FwdIt：前向迭代器。
- T：比较值的类型。
- Pred：二元判定函数。

该函数功能是：容器元素已经排好序，在[0，last−first)范围内寻找位置 N，M∈[0,N)。对第一个模板函数而言，*(first+M) < val 是 true，*(first+N)≥val，也就是说，在有序容器中寻找第一个大于等于 val 值的位置，若找到返回 first+N，否则返回 last；对第二个模板函数而言，功能相似，返回第一个不满足 pr(*(first+M)，val)的位置 N。

② upper_bound

原型：

```
template<class FwdIt, class T>
    FwdIt upper_bound(FwdIt first, FwdIt last, const T& val);
template<class FwdIt, class T, class Pred>
    FwdIt upper_bound(FwdIt first, FwdIt last, const T& val, Pred pr);
```

参数说明：
- FwdIt：前向迭代器。
- T：比较值的类型。
- Pred：二元判定函数。

该函数功能是：容器元素已经排好序，在[0，last－first)范围内寻找位置 N，M∈[0,N)。对第一个模板函数而言，*(first＋M)≤val 是 true，*(first＋N)＞val，也就是说，在有序容器中寻找第一个大于 val 值的位置，若找到返回 first＋N，否则返回 last；对第二个模板函数而言，功能相似，返回第一个不满足 pr(*(first＋M)，val)的位置 N。

③ equal_range

原型：

```
template<class FwdIt, class T>
    pair<FwdIt, FwdIt>equal_range(FwdIt first, FwdIt last,const T& val);
template<class FwdIt, class T, class Pred>
    pair<FwdIt, FwdIt>equal_range(FwdIt first, FwdIt last,const T& val, Pred pr);
```

参数说明：
- FwdIt：前向迭代器。
- T：比较值的类型。
- Pred：二元判定函数。

第一个模板函数功能是：在有序元素容器中，找出一对迭代指针 midstart，midend，其间的元素都等于 val，即 N∈[0,midend－midstart)，*(midstart＋N)＝val。如果有结果，结果保存在 pair 对象中，相当于 pair(lower(first，last，val)，upper(first,last,val))。第二个模板函数与第一个模板函数功能相近：找出一对迭代指针 midstart，midend，其间元素：N∈[0,midend－midstart)，pr(*(midstart＋N)，val)都是 true。

④ binary_search

原型：

```
template<class FwdIt, class T>
    bool binary_search(FwdIt first, FwdIt last, const T& val);
template<class FwdIt, class T, class Pred>
    bool binary_search(FwdIt first, FwdIt last, const T& val,Pred pr);
```

参数说明：
- FwdIt：前向迭代器。
- T：比较值的类型。
- Pred：二元判定函数。

第一个模板函数功能是：在有序容器中查询*[first,last)间有无元素值等于 val，若有返回 true，若无返回 false；第二个模板函数功能是：在有序容器中查询*[first,last)间有无元素值满足 N∈[0,last－first)，若 pr(*(first＋N)，val)成立则返回 true，若无返回 false。

10.3.2 示例分析

【例 10.7】 已知 list 有序容器{1,2,2,3,3,3,4,4,4,4},求:(1)有无元素 5?(2)第一个元素 2 的位置;(3)最后一个元素 3 的位置;(4)共有多少个元素 4?

```cpp
//文件名:e10_7.cpp
#include <iostream>
#include <algorithm>
#include <list>
#include <iterator>
using namespace std;
int main()
{
    int a[] ={1,2,2,3,3,3,4,4,4,4};
    list<int>l1(a, a+10);
    cout<<"原始数据 a[]:";
    copy(l1.begin(), l1.end(), ostream_iterator<int>(cout, "\t"));
    cout<<endl;

    bool bExist =binary_search(l1.begin(), l1.end(), 5);
    cout<<"有元素 5 吗: "<< (bExist==0?"false":"true")<<endl;

    list<int>::iterator first2 =lower_bound(l1.begin(), l1.end(),2);
                                                        //求第一个元素 2 的位置
    if(first2 !=l1.end())                               //找到第一个元素 2 的位置
    {
        cout<<"第一个元素 2 的位置:"<<distance(l1.begin(), first2)<<endl;
    }

    list<int>::iterator last3 =upper_bound(l1.begin(), l1.end(),3);
                                                        //最后一个元素 3 之后的指针
    if(last3 !=l1.end())                                //找到最后一个元素 3 的位置
    {
        cout<<"最后一个元素 3 的位置:"<<distance(l1.begin(), --last3)<<endl;
    }

    pair<list<int>::iterator,list<int>::iterator>p=equal_range(l1.begin(), l1.end(), 4);
    if(p.first!=p.second)
    {
        int nSize =distance(first4, last4);
        cout<<"共有元素 4 个数:"<<nSize<<endl;
```

        }
        return 0;
}

执行结果是：

原始数据 a[]: 1 2 2 3 3 3 4 4 4 4
有元素 5 吗: false
第一个元素 2 的位置:1
最后一个元素 3 的位置:5
共有元素 4 个数: 4

(1) 若想在有序容器中查找有无某值，用 binary_search 函数，它仅返回 bool 值，并不能返回查找结果的迭代器指针；若想在有序容器中查找某值位置，要用到 lower_bound, upper_bound 函数。如果结果有效，则 lower_bound 返回迭代指针指向的元素正好是某值，而 upper_bound 返回迭代指针正好是最后一个某值的下一个迭代指针。因此 lower_bound, upper_bound 函数的返回值含义是不一样的，所以当计算第一个某值位置与计算最后一个某值位置的方法稍有差别。本例中，假设元素位置从 0 开始，第一个元素 2 位置代码为 distance(ll.begin(), first2)，最后一个元素 3 位置代码为 distance(ll.begin(), --last3)。若想在有序容器中求共有几个某值，则直接应用 equal_range 即可。

(2) 可以看出，根据查询语义选择函数非常重要。一般来说，若在有序元素内，则选择本节所讲的 4 个二分函数，与容器元素规模相比，仅需要对数时间；若在无序元素内，则选用非变异 find 系列函数等，与容器元素规模相比，它需要线性时间。

10.4 归并

10.4.1 主要函数

- merge()：归并双迭代器两个有序序列。
- inplace_merge()：归并单迭代器两个接续的有序序列。

函数原型如下所示。

① merge

原型：

template<class InIt1, class InIt2, class OutIt>
  OutIt **merge**(InIt1 first1, InIt1 last1,InIt2 first2, InIt2 last2, OutIt x);
template<class InIt1, class InIt2, class OutIt, class Pred>
  OutIt **merge**(InIt1 first1, InIt1 last1,InIt2 first2, InIt2 last2, OutIt x, Pred pr);

参数说明：

- InIt1，InIt2：输入迭代器。

- OutIt：输出迭代器。
- Pred：二元函数。

第一个模板函数功能是：两个有序容器元素（均按 operator＜排序）交替比较，按 operate＜排序后，输出至 x 迭代器表示的容器中。另外，当两个容器中比较元素相同，则第一个容器中元素先输出到 x 代表的容器中。若两个容器共有 N 个元素合并，则返回 x＋N。第二个函数与第一个函数功能相近，只是用二元函数 pr 代替了 operate＜。

② inplace_merge

原型：

```
template<class BidIt>
  void inplace_merge(BidIt first, BidIt middle, BidIt last);
template<class BidIt, class Pred>
  void inplace_merge(BidIt first, BidIt middle, BidIt last, Pred pr);
```

参数说明：
- BidIt：双向迭代器。
- Pred：二元函数。

第一个模板函数功能是：一个容器分为两部分[first,middle)、[middle,last)，每部分都已按 operator＜排好序，但整体不一定排好序。当进行 inplace_merge 时，[first,middle)中指向元素交替与[middle,last)中指向元素按 operate＜排序后，放入原容器中。当比较元素相同时，[first,middle)间元素优先存放。第二个函数与第一个函数功能相近，只是用二元函数 pr 代替了 operate＜。

10.4.2 示例分析

【例 10.8】 合并函数简单示例。

```cpp
//文件名：e10_8.cpp
#include <iostream>
#include <algorithm>
#include <iterator>
using namespace std ;
int main()
{
    int a[]={1,3,5,7};
    int b[]={2,4,6,8};
    int result[8];
    cout<<"原始 a[]:"<<endl;
    copy(a, a+4, ostream_iterator<int>(cout, "\t"));           //1 3 5 7
    cout<<endl;
    cout<<"原始 b[]:" <<endl;
```

```
        copy(b, b+4, ostream_iterator<int>(cout, "\t"));        //2 4 6 8
        cout<<endl;
        merge(a, a+4, b, b+4, result);
        cout<<"a[],b[]merge 后:"<<endl;
        copy(result, result+8, ostream_iterator<int>(cout,"\t"));    //1 2 3 4 5 6 7 8
        cout<<endl<<endl;

        int c[8]={1,3,4,8,2,5,6,7 };
        cout<<"原始 c[]:"<<endl;
        copy(c, c+8, ostream_iterator<int>(cout, "\t"));            //1 3 4 8 2 5 6 7
        cout<<endl;
        inplace_merge(c, c+4, c+8);
        cout<<"c[] inplace_merge 后"<<endl;
        copy(c, c+8, ostream_iterator<int>(cout, "\t"));//1 2 3 4 5 6 7 8
        cout<<endl;
        return 0;
}
```

(1) a[]、b[]均是升序,合并排序后结果仍是升序,但是要注意:不能一个升序、一个降序。例如 a[]={1,3,5,7},b[]={8,6,4,2},虽然 merge 后也有结果,但它只不过执行了一遍 merge 函数,结果就不一定正确了。同理,inplace_merge 要求两部分元素都有序,而且要么升序,要么降序。本例中以 c[4]划分,可看出前部分{1 3 4 8},后部分{2 4 6 8}都是升序,所以可得正确的合并结果。

(2) 假如 a[]={7,5,3,1},b[]={8,6,4,2},现在需要合并后仍然降序排列,下述代码可以吗?

```
int a[]={7,5,3,1}; int b[]={8,6,4,2}; int result[8];
merge(a, b, result);
```

因为 a[]、b[]都是按 operator>排好序的,采用的是 merge 模板函数的第一种形式,它是按 operator<合并排序的,前后操作符不统一,结果一定是错误的。应该用第二个模板函数,代码如下:

```
merge(a, b, result, greater<int>());
```

10.5 有序结构上的集合操作

10.5.1 主要函数

- includes():一个序列为另一个序列的子序列时为真。
- set_union():构造两个集合的有序并集。

- set_intersection()：构造两个集合的有序交集。
- set_difference()：构造两个集合的有序差集。
- set_symmetric_difference()：构造两个集合的有序对称差集(并—交)。

函数原型如下。

① includes

原型：

```
template<class InIt1, class InIt2>
    bool includes(InIt1 first1, InIt1 last1,InIt2 first2, InIt2 last2);
template<class InIt1, class InIt2, class Pred>
    bool includes(InIt1 first1, InIt1 last1,InIt2 first2, InIt2 last2, Pred pr);
```

参数说明：

- InIt1,InIt2：输入迭代器。
- Pred：二元函数。

第一个模板函数功能是：两个容器按 operate< 已排好序,若容器[first2,last2)指向的每个元素都在[first1,last1)指向的元素范围内,则[first1,last1)代表容器包含[first2,last2)代表的容器。第二个模板函数功能与第一个功能相近,两个容器按 pr 已排好序,若容器[first2,last2)指向的每个元素都在[first1,last1)指向的元素范围内,则[first1,last1)代表容器包含[first2,last2)代表的容器。

② set_union

原型：

```
template<class InIt1, class InIt2, class OutIt>
    OutIt set_union(InIt1 first1, InIt1 last1,InIt2 first2, InIt2 last2, OutIt x);
template<class InIt1, class InIt2, class OutIt, class Pred>
    OutIt set_union(InIt1 first1, InIt1 last1,InIt2 first2, InIt2 last2, OutIt x, Pred pr);
```

参数说明：

- InIt1,InIt2：输入迭代器。
- OutIt：输出迭代器。
- Pred：二元函数。

第一个模板函数功能是：两个容器[first1,last1)、[first2,last2)均按 operate< 排好序,两个容器元素交替比较,小的值进入输出容器 x 中。若比较两值相等,则取[first1,last1)中的元素。如果有 N 个元素复制到 x 代表的容器中,则返回 x+N。第二个模板函数功能与第一个函数功能相近,只不过用二元函数 pr 代替了 operate<。

③ set_intersection

原型：

```
template<class InIt1, class InIt2, class OutIt>
    OutIt set_intersection(InIt1 first1, InIt1 last1,InIt2 first2, InIt2 last2,
```

```
    OutIt x);
template<class InIt1, class InIt2, class OutIt, class Pred>
    OutIt set_intersection(InIt1 first1, InIt1 last1,InIt2 first2, InIt2 last2,
    OutIt x, Pred pr);
```

参数说明:
- InIt1, InIt2: 输入迭代器。
- OutIt: 输出迭代器。
- Pred: 二元函数。

第一个模板函数功能是:两个容器[first1,last1)、[first2,last2)均按 operate< 排好序,两个容器元素交替比较,若比较两值相等,则取[first1,last1)中的元素。如果有 N 个元素拷贝到 x 代表的容器中,则返回 x+N。第二个模板函数功能与第一个函数功能相近,只不过用二元函数 pr 代替了 operate<。

④ set_difference

原型:

```
template< class InIt1, class InIt2, class OutIt>
    OutIt set_difference(InIt1 first1, InIt1 last1,InIt2 first2, InIt2 last2,
    OutIt x);
template<class InIt1, class InIt2, class OutIt, class Pred>
    OutIt set_difference(InIt1 first1, InIt1 last1,InIt2 first2, InIt2 last2, OutIt x,
    Pred pr);
```

参数说明:
- InIt1, InIt2: 输入迭代器。
- OutIt: 输出迭代器。
- Pred: 二元函数。

第一个模板函数功能是:两个容器[first1,last1)、[first2,last2)均按 operate< 排好序,两个容器元素交替比较,若比较两值相等,则取[first1,last1)中的元素。如果有 N 个元素拷贝到 x 代表的容器中,则返回 x+N。第二个模板函数功能与第一个函数功能相近,只不过用二元函数 pr 代替了 operate<。

⑤ set_symmetric_difference

原型:

```
template<class InIt1, class InIt2, class OutIt>
    OutIt set_symmetric_difference(InIt1 first1, InIt1 last1, InIt2 first2, InIt2
    last2, OutIt x);
template< class InIt1, class InIt2, class OutIt, class Pred>
    OutIt set_symmetric_difference(InIt1 first1, InIt1 last1, InIt2 first2, InIt2
    last2, OutIt x, Pred pr);
```

参数说明:

- InIt1，InIt2：输入迭代器。
- OutIt：输出迭代器。
- Pred：二元函数。

10.5.2 示例分析

【例 10.9】 集合操作简单示例。

```
//文件名：e10_9.cpp
#include <iostream>
#include <list>
#include <algorithm>
#include <iterator>
using namespace std ;
int main()
{
    int a[]={1,2,3,4,5};
    int b[]={1,2,3,6};

    list<int>l1(a, a+5);
    list<int>l2(b, b+4);

    cout<<"原始 l1:";
    copy(l1.begin(), l1.end(), ostream_iterator<int>(cout,"\t"));
    cout<<endl;
    cout<<"原始 l2:";
    copy(l2.begin(), l2.end(), ostream_iterator<int>(cout,"\t"));
    cout<<endl;
    //包含
    bool bRet=includes(l1.begin(), l1.end(), l2.begin(),l2.end());
    cout<<"l1 包含 l2?" << (bRet==1?"yes":"no" )<<endl;
    //l1 并 l2
    list<int>l3;
    set_union(l1.begin(),l1.end(), l2.begin(),l2.end(), back_inserter(l3));
    cout<<"l1 并 l2:";
    copy(l3.begin(), l3.end(), ostream_iterator<int>(cout,"\t"));
    cout<<endl;
    //l1 交 l2
    list<int>l4;
    set_intersection(l1.begin(), l1.end(), l2.begin(), l2.end(), back_inserter(l4));
    cout<<"l1 交 l2:";
    copy(l4.begin(), l4.end(), ostream_iterator<int>(cout,"\t"));
```

```cpp
        cout<<endl;
        //l1差l2
        list<int>l5;
        set_difference(l1.begin(), l1.end(), l2.begin(), l2.end(), back_inserter(l5));
        cout<<"l1差l2:";
        copy(l5.begin(), l5.end(), ostream_iterator<int>(cout,"\t"));
        cout<<endl;
        //l1对称差l2
        list<int>l6;
        set_symmetric_difference(l1.begin(), l1.end(), l2.begin(), l2.end(), back_inserter(l6));
        cout<<"l1对称差l2:";
        copy(l6.begin(), l6.end(), ostream_iterator<int>(cout,"\t"));
        cout<<endl;
        //l1 merge l2
        list<int>l7;
        merge(l1.begin(), l1.end(), l2.begin(), l2.end(), back_inserter(l7));
        cout<<"l1 merge l2:";
        copy(l7.begin(), l7.end(), ostream_iterator<int>(cout,"\t"));
        cout<<endl;
        return 0;
    }
```

执行结果是：

原始 l1:1 2 3 4 5
原始 l2:1 2 3 6
l1 包含 l2? No
l1 并 l2:1 2 3 4 5 6
l1 交 l2:1 2 3
l1 差 l2:4 5
l1 对称差 l2: 4 5 6
l1 merge l2:1 1 2 2 3 3 4 5 6

(1) l1+l2,l1－l2,l1∩l2 都比较容易理解,强调一下"l1 对称差"相当于(l1－l2)＋(l2－l1)或(l1＋l2)－(l1∩l2)。

(2) "l1 并 l2"与"l1 merge l2"结果是不同的,可以看出 set_union 函数与 merge 函数的区别。两者都具有合并功能,但 merge 保留重复的元素,而 set_union 去掉重复元素。或者这样理解：若两个容器有待合并元素个数 n1、n2,则 merge 后元素个数一定是 n1＋n2,而 set_union 后元素个数小于等于 n1＋n2。

(3) 假如 a[]、b[]降序排列,a[]＝{5,4,3,2,1},b[]＝{3,2,1},如果判定 a 是否包含 b,下述代码是错误的：includes(a,a+5,b,b+3)。这是由于 a[]、b[]本身是按 operator>排序得到的,而 includes(a,a+5,b,b+3)是按 operator<判定是否是包含关系,操作符不统

一,所以是错误的,只要用 includes 第二个模板函数就可以了,即 includes(a,a+5, b,b+3, greater<int>())。

【例 10.10】 深入理解序结构集合操作中 operator< 函数作用。已知两个学生集合初始是无序的,要求按学号排序后判定第一个学生集合是否包含第二个学生集合。

```cpp
//文件名:e10_10.cpp
#include <iostream>
#include <algorithm>
using namespace std ;
struct Student
{
    int NO;                          //学号
    char name[20];                   //姓名
    bool operator< (const Student& s)const
    {
        return NO< s.NO;
    }
};
int main()
{
    bool bRet=true;
    Student s1[]={{1001,"zhang"},{1003,"li"},
                  {1004,"wang"},{1002,"zhao"}};      //第一个学生模拟集合
    Student s2[]={{1001,"zhang"},{1004,"wang"},
                  {1002,"zhao"}};                    //第二个学生模拟集合
    sort(s1, s1+4);                                  //排序
    sort(s2, s2+3);                                  //排序
    bRet=includes(s1, s1+4, s2, s2+3);               //判定包含关系
    cout<< (bRet==true?"s1 包含 s2":"s1 不包含 s2")<<endl;
    return 0;
}
```

可以看出: operator< 函数在 sort 函数中起作用,也在 includes 中起作用。sorts 函数要完成比较功能,重载 operator< 是容易理解的。根据 includes 语义,总感觉一方面需要比较,需重载 operator<;另一方面需要比较是否相等,需重载 operator==。但是根本没有重载 operator==。其实本节 5 个集合函数第一个模板函数都仅需重载 operator< 即可。先看一下 includes 函数源码:

```cpp
template<class _II1, class _II2>inline
bool includes(_II1 _F1, _II1 _L1, _II2 _F2, _II2 _L2)
{
    for (; _F1 != _L1 && _F2 != _L2; )
        if (*_F2 < *_F1)                //*F1< *F2
```

```
            return (false);
        else if (* _F1 < * _F2)         // * F2< * F1
            ++ _F1;
        else                             // * F1== * F2
            ++ _F2;
    return (_F2== _L2);
}
```

* _F1 和 * _F2 有三种关系：>、=、<。若 * _F1 大于 * _F2,代码用" * _F2 < * _F1"表示；若 * _F1 小于 * _F2,代码用" * _F1 < * _F2"表示；若 * _F1 等于 * _F2,代码中并没有用" * _F1== * _F2"表示，而是巧妙地采用了 if-else if-else 结构,最后的 else 表明了" * _F1== * _F2"。所以在 includes 中仅需重载 operator< 即可。

【例 10.11】 已知学生结构体包含信息：学号(int)、姓名(string)、语文成绩(int)、数学成绩(int)。语文老师形成了一个语文成绩文件,数学老师形成了一个数学成绩文件,均已按学号升序排好序。把学生语文成绩与数学成绩合并,形成学生的全信息,如表 10.2 所示。要考虑某学生某科缺考情况(按 0 分处理)。

表 10.2　语文、数学成绩文件说明

语文老师成绩文件 d:\chinese.txt				数学老师成绩文件 d:\math.txt			
文本格式：	学号	姓名	语文成绩	文本格式：	学号	姓名	数学成绩
	1001	zhang	75		1001	zhang	77
	1002	li	85		1002	li	87
	1003	sun	78		1003	sun	80
					1004	zhao	85

表 10.2 表明：zhang,li,sun 考了语文、数学；zhao 考了数学,语文缺考。

```
//文件名：e10_11.cpp
#include <iostream>
#include <fstream>
#include <string>
#include <vector>
#include <algorithm>
using namespace std;
struct Student
{
    int NO;
    string name;
    int chinese;
    int math;
public:
    bool operator< (const Student & s)
```

```cpp
        if(NO==s.NO)
        {
            math=s.math;
        }
        return NO<s.NO;
    }
};
int main(int argc, char * argv[])
{
    vector<Student>v1;                              //语文成绩向量
    vector<Student>v2;                              //数学成绩向量

    ifstream in1("d:\\chinese.txt");                //打开语文成绩文件
    ifstream in2("d:\\math.txt");                   //打开数学成绩文件
    while(!in1.eof())                               //读语文成绩文件并存入向量
    {
        Student s;
        in1>>s.NO>>s.name>>s.chinese;
        v1.push_back(s);
    }
    while(!in2.eof())                               //读数学成绩文件并存入向量
    {
        Student s;
        in2>>s.NO>>s.name>>s.math;
        v2.push_back(s);
    }

    in1.close();
    in1.close();

    set_union(v1.begin(), v1.end(), v2.begin(), v2.end(), v1.begin());   //合并
    for(int i=0; i<v1.size(); i++)
    {
        Student & s=v1.at(i);
        cout<<s.NO<<"\t"<<s.name<<"\t"<<s.chinese<<"\t"<<s.math<<endl;
    }

    return 0;
}
```

语文成绩文件与数学成绩文件是两个不同的文件,如何把它们合并在一起是关键所在。两个文件中学号和姓名是重复项,合并的时候要去掉重复项,因此应该选择 set_union 函

数,要把语文成绩向量 v1 与数学成绩向量 v2 进行 set_union 操作,但是语文成绩项与数学成绩项分别在两个向量中,而 set_union 通过 operator<操作只能保留 v1 向量或 v2 向量的一个元素,感觉上几乎不可能完成语文、数学成绩的合并。其实只要在基础类重载的 operator<函数中实现一个小技巧就可以了,把代码再写在下面加以分析:

```
bool operator<(const Student&s)
{
    if(NO==s.NO)
    {
        math=s.math;
    }
    return NO<s.NO;
}
```

着重注意黑体部分代码:当前对象表明该学生具有语文成绩,传入的函数参数 s 代表传入待比较的学生对象具有数学成绩,如果当前学号等于传入对象的学号,则说明是同一个学生,则把传入学生对象的数学成绩赋给当前对象的数学变量 math,在 operator<函数中巧妙地实现了对象成员变量的修改。因此 set_union 函数不断地调用 operator<函数,一方面消除了语文成绩向量与数学成绩向量重复的部分,另一方面又恰当地修改了结果向量的值。需要同学们好好加以思考。

10.6 堆操作

如果有一个关键码的集合 $K=\{k_0,k_1,k_2,\cdots,k_n\}$,把它的所有元素按完全二叉树的顺序存储方式存放在一个一维数组中,并且满足 $k_i \leqslant k_{2i+1}$ 且 $k_i \leqslant k_{2i+2}$ (或 $k_i \geqslant k_{2i+1}$ 且 $k_i \geqslant k_{2i+2}$),则称这个集合为最小堆(或者最大堆)。

10.6.1 主要函数

- make_heap():从序列构造堆。
- pop_heap():从堆中弹出元素。
- push_heap():向堆中加入元素。
- sort_heap():给堆排序。

函数原型如下所示。

① make_heap

原型:

```
template<class RanIt>
    void make_heap(RanIt first, RanIt last);
```

```
template<class RanIt, class Pred>
    void make_heap(RanIt first, RanIt last, Pred pr);
```

参数说明：
- RanIt：随机迭代器。
- Pred：二元函数。

第一个模板函数功能是：把随机迭代器[first, last)间元素按 operator< 排序，形成一个堆。第二个函数与第一个函数功能相近，只不过用二元函数 pr 代替了 operator<。

② pop_heap

原型：

```
template<class RanIt>
    void pop_heap(RanIt first, RanIt last);
template<class RanIt, class Pred>
    void pop_heap(RanIt first, RanIt last, Pred pr);
```

参数说明：
- RanIt：随机迭代器。
- Pred：二元函数。

第一个模板函数功能是：不是真的把最大（最小）的元素从堆中弹出来，而是重新排序堆。它先将 first 和 last 交换，然后将[first, last-1)的数据按 operator< 再做成一个堆。第二个函数与第一个函数功能相近，只不过用二元函数 pr 代替了 operator<。

③ push_heap

原型：

```
template<class RanIt>
    void push_heap(RanIt first, RanIt last);
template<class RanIt, class Pred>
    void push_heap(RanIt first, RanIt last, Pred pr);
```

参数说明：
- RanIt：随机迭代器。
- Pred：二元函数。

第一个模板函数功能是：假设[first, last-1)迭代器对应元素是一个有效堆，然后再把堆中的新元素加进来，按 operator< 做成一个堆。第二个函数与第一个函数功能相近，只不过用二元函数 pr 代替了 operator<。

④ sort_heap

原型：

```
template<class RanIt>
    void sort_heap(RanIt first, RanIt last);
template<class RanIt, class Pred>
```

```
void sort_heap(RanIt first, RanIt last, Pred pr);
```

参数说明:
- RanIt: 随机迭代器。
- Pred: 二元函数。

第一个模板函数功能是: 对[first,last)中的序列按 operator<进行排序,它假设这个序列是有效堆。当然,经过排序之后就不是一个有效堆了。第二个函数与第一个函数功能相近,只不过用二元函数 pr 代替了 operator<。

10.6.2 示例分析

【例 10.12】 已知整型序列{1,4,2,10,6,5,9,7,8,3},(1)建最大堆;(2)建最小堆;(3)求堆中最大值及次大值。

```cpp
//文件名: e10_12.cpp
#include<iostream>
#include<vector>
#include<functional>
#include<algorithm>
#include<iterator>
using namespace std;
int main()
{
    int a[]={1,4,2,10,6,5,9,7,8,3};
    cout<<"原始 a[]:";
    copy(a, a+10, ostream_iterator<int>(cout, "\t"));
    cout<<endl;

    vector<int>v1(a, a+10);
    make_heap(v1.begin(), v1.end(), greater<int>());        //形成最小堆
    cout<<"最小堆:";
    copy(v1.begin(), v1.end(), ostream_iterator<int>(cout, "\t"));
    cout<<endl;

    vector<int>v2(a, a+10);
    make_heap(v2.begin(), v2.end(), less<int>());           //形成最大堆
    cout<<"最大堆:";
    copy(v2.begin(), v2.end(), ostream_iterator<int>(cout, "\t"));
    cout<<endl;

    pop_heap(v2.begin(), v2.end());
    cout<<"堆中最大值:";
```

```
        cout<< * (v2.end()-1)<<endl;
        pop_heap(v2.begin(), v2.end()-1);
        cout<<"堆中次大值:";
        cout<< * (v2.end()-2)<<endl;
        return 0;
    }
```

(1) 建最小堆用的二元函数是 greater,不是 less;建最大堆用的二元函数是 less,而不是 greater。与常规用到的二元函数语义正好相反。

(2) 若向量容器 v 已建好最大堆,求最大元素应执行 pop_heap(v.begin(), v.end()),这样[v.begin(),v.end()-1]仍是有效堆,而最大值存放在 *(e.end()-1)中;若求次大值,应执行 pop_heap(v.begin(), v.end()-1),这样[v.begin(),v.end()-2]仍是有效堆,而最大值存放在 *(e.end()-2)中。往下以此类推。

【例 10.13】 进一步封装 4 个堆操作函数,并编制测试类加以测试。

```
//文件名:e10_13.cpp
#include <iostream>
#include <vector>
#include <functional>
#include <algorithm>
#include <iterator>
using namespace std;
template<class T,class Pred=less<T>>
class MyHeap
{
    Pred pr;
    vector<T>v;
    int ValidSize;
public:
    MyHeap(T t[], int nSize):v(t, t+nSize)                          //建立堆
    {
        ValidSize=nSize;
        make_heap(v.begin(), v.begin()+ValidSize, pr);
    }
    void Insert(const T & t)                                        //向堆中新加元素
    {
        v.push_back(t);
        ValidSize++;
        push_heap(v.begin(), v.begin()+ValidSize, pr);
    }
    bool Remove(T& result)                                          //获得堆中最大值或最小值
    {
        if(ValidSize==0)
```

```cpp
            return false;
        pop_heap(v.begin(), v.begin()+ValidSize, pr);
        result= * (v.begin()+ValidSize-1);
        ValidSize--;
        return true;
    }
    void Sort()
    {
        if(ValidSize>0)
            sort_heap(v.begin(), v.begin()+ValidSize);
    }
    bool IsEmpty()                                    //判断堆是否为空
    {
        return ValidSize==0;
    }
    void Display()
    {
        copy(v.begin(), v.begin()+ValidSize, ostream_iterator<T>(cout, "\t"));
    }
};
```

（1）模板参数 T 表示堆中元素类型，二元函数默认是 less<T>，表明默认生成一个最大堆的排列。若生成一个最小堆排列，只需把 greater<T>作为模板参数传进去就可以了。

（2）成员变量 ValidSize 表明当前堆的有效大小，也就是说[v. begin()，v. begin()+ValidSize)指向的元素序列是一个有效堆。

（3）构造函数是传入一个数组，直接调用 make_heap 形成了一个有效堆。当然，构造函数还可以进行重载，比如参数是 vector，同学们可以加以扩充。

（4）4 个主要的函数：Insert，完成堆元素添加，同时堆有效长度 ValidSize＋＋；Remove，获得堆中最大值（或最小值），同时堆有效长度 ValidSize－－；Sort，堆排序函数；IsEmpty，根据 ValidSize 判断堆是否为空。

用下面代码加以测试，可以看出经过封装后，调用形式简化了许多。

```cpp
int main()
{
    int a[]={1,4,2,10,6};
    MyHeap<int,greater<int>>m(a, 5);              //产生一个最小堆
    cout<<"原始堆:"<<endl;
    m.Display(); cout<<endl;                       //1 4 2 10 6
    m.Insert(100);
    cout<<"插入 100 后堆变为:"<<endl;
    m.Display(); cout<<endl;                       //1 4 2 10 6 100
```

```
        int result;
        m.Remove(result);
        cout<<"堆中最小值:"<<result<<endl;             //1
        m.Remove(result);
        cout<<"堆中次小值:"<<result<<endl;             //2
        return 0;
    }
```

10.7 最大和最小

10.7.1 主要函数

- min()：两个值中较小的。
- max()：两个值中较大的。
- min_element()：序列中的最小元素。
- max_element()：序列中的最大元素。

函数原型如下所示。

① min

原型：

```
template<class T>
    const T& min(const T& x, const T& y);
template<class T, class Pred>
    const T& min(const T& x, const T& y, Pred pr);
```

参数说明：

- T：元素类型。
- Pred：二元比较函数。

第一个模板函数功能是：按 operator< 返回两个元素中较小的一个。第二个函数与第一个函数功能相近，只不过用二元比较函数 pr 代替了 operator<。

② max

原型：

```
template<class T>
    const T & max(const T & x, const T & y);
template<class T, class Pred>
    const T & max(const T & x, const T & y, Pred pr);
```

参数说明：

- T：元素类型。

- Pred：二元比较函数。

第一个模板函数功能是：按 operator< 返回两个元素中较大的一个。第二个函数与第一个函数功能相近，只不过用二元比较函数 pr 代替了 operator<。

③ min_element

原型：

template< class FwdIt>
　FwdIt **min_element**(FwdIt first, FwdIt last);
template< class FwdIt, class Pred>
　FwdIt **min_element**(FwdIt first, FwdIt last, Pred pr);

参数说明：
- FwdIt：前向迭代器。
- Pred：二元比较函数。

第一个模板函数功能是：按 operator< 比较，找出[first,last)指向元素的最小值，若设第 N 个元素是最小值，则返回 first+N。第二个函数与第一个函数功能相近，只不过用二元比较函数 pr 代替了 operator<。

④ max_element

原型：

template<class FwdIt>
　FwdIt **max_element**(FwdIt first, FwdIt last);
template<class FwdIt, class Pred>
　FwdIt **max_element**(FwdIt first, FwdIt last, Pred pr);

参数说明：
- FwdIt：前向迭代器。
- Pred：二元比较函数。

第一个模板函数功能是：按 operator< 比较，找出[first,last)指向元素的最大值，若设第 N 个元素是最大值，则返回 first+N。第二个函数与第一个函数功能相近，只不过用二元比较函数 pr 代替了 operator<。

10.7.2　示例分析

【例 10.14】　最大和最小函数简单示例。

```
//文件名：e10_14.cpp
#include <iostream>
#include <algorithm>
using namespace std;

int main()
```

```
{
    int a=min(10,20);
    int b=max(10,20);
    cout<<"min(10,20):"<<a<<endl;
    cout<<"max(10,20):"<<b<<endl;

    int c[]={1,10,5,7,9};
    cout<<"原始c[]:";
    copy(c, c+5, ostream_iterator<int>(cout, "\t"));          //1 10 5 7 9
    cout<<endl;
    int * it_min=min_element(c, c+5);
    int * it_max=max_element(c, c+5);
    cout<<"c[]最小值:"<< * it_min<<endl;                      //1
    cout<<"c[]最大值:"<< * it_max<<endl;                      //10
    return 0;
}
```

该示例很简单,由于 windef.h 中也定义了 min、max 函数,为了避免冲突,用宏_MIN、_MAX 代替了 min、max。

10.8 词典比较

10.8.1 主要函数

lexicographical_compare():两个序列按字典序排序。
函数原型如下所示。

```
template<class InIt1,class InIt2>
  bool lexicographical_compare(InIt1 first1, InIt1 last1,InIt2 first2,
    InIt2 last2);
template<class InIt1, class InIt2, class Pred>
  bool lexicographical_compare(InIt1 first1, InIt1 last1,InIt2 first2, InIt2 last2,
    Pred pr);
```

参数说明:
- InIt1,InIt2:输入迭代器。
- Pred:二元比较函数。

第一个模板函数功能是:该算法比较两个序列中的对应元素 e1 和 e2(分别来自序列 1 和序列 2)。如果 e1<e2,则算法立即返回,返回值为真;如果 e2<e1,则算法也立即返回,返回值为假;否则继续对下一对元素进行比较。如果已到达第一个序列的末尾,但没有到达第二个序列的末尾,则算法返回,返回值为真;否则返回假。第二个函数与第一个函数

功能相近,只不过用二元比较函数 pr 代替了 operator<。

10.8.2 示例分析

【例 10.15】 lexicographical_compare 词典比较函数简单示例。

```
//文件名: e10_15.cpp
int main()
{
    int a[]={1,2,3};
    int b[]={1,2,2};
    bool bRet=lexicographical_compare(a, a+3, b, b+3);
    cout<< (bRet==1?"a[]<b[]":"a[]>b[]")<<endl; //a[]>b[]
    return 0;
}
```

该示例很简单,但并不能看出 lexicographical_compare 有什么用处。再看一个集合排序问题。例如假设 int a[]={1,2,3}, int b[]={1,2,2}, int c[]={1,2,3,1},如果想对这三个容器排序,就需要在容器间加以比较,这时候就可能用到 lexicographical_compare 函数了,代码如下:

```
bool mycompare(vector<int>& v1, vector<int>& v2)
{
    return lexicographical_compare(v1.begin(), v1.end(), v2.begin(), v2.end());
}
void main()
{
    int a[]={1,2,3};
    int b[]={1,2,2};
    int c[]={1,2,3,1};
    vector<int>v[3]={vector<int>(a,a+3),vector<int>(b,b+3),vector<int>(c,c+4)};
    cout<<"原始 3 个向量:"<<endl;
    for(int i=0; i<3; i++)
    {
        copy(v[i].begin(), v[i].end(), ostream_iterator<int>(cout, "\t"));
        cout<<endl;
    }
    sort(v, v+3, mycompare);
    cout<<"排序后(按 lexicographical_compare 比较)3 个向量:"<<endl;
    for(i=0; i<3; i++)
    {
        copy(v[i].begin(), v[i].end(), ostream_iterator<int>(cout, "\t"));
        cout<<endl;
```

    }
}

在 mycompare 自定义函数中，传入参数是两个容器，两个容器按词典序比较直接调用 lexicographical_compare 函数就可以了。如果没有该函数，那么就要写较复杂的代码。因此 lexicographical_compare 是按词典序比较的函数框架，可以方便加以扩展。

10.9 排列生成器

10.9.1 主要函数

- next_permutation()：按字典序的下一个排列。
- prev_permutation()：按字典序的上一个排列。

函数原型如下。

① next_permutation

原型：

```
template<class BidIt>
  bool next_permutation(BidIt first,BidIt last);
template<class BidIt, class Pred>
  bool next_permutation(BidIt first, BidIt last,Pred pr)
```

参数说明：

- BidIt：双向迭代器。
- Pred：二元比较函数。

第一个模板函数功能是：按 operator<生成[first，last)指向容器的下一个词典序排列。第二个函数与第一个函数功能相近，只不过用二元比较函数 pr 代替了 operator<。

② prev_permutation

原型：

```
template <class BidIt>
  bool prev_permutation(BidIt first,BidIt last);
template <class BidIt, class Pred>
  bool prev_permutation(BidIt first,BidIt last,Pred pr)
```

参数说明：

- BidIt：双向迭代器。
- Pred：二元比较函数。

第一个模板函数功能是：按 operator<生成[first，last)指向容器的上一个词典序排列。第二个函数与第一个函数功能相近，只不过用二元比较函数 pr 代替了 operator<。

10.9.2 示例分析

【例 10.16】 排列生成器简单示例。

```cpp
//文件名：e10_16.cpp
#include <iostream>
#include <algorithm>
#include <iterator>
using namespace std;
int main()
{
  int A[]={2, 3, 4, 5, 6, 1};
  const int N=sizeof(A)/sizeof(int);

  cout<<"初始化：";
  copy(A, A+N, ostream_iterator<int>(cout, " "));//2 3 4 5 6 1
  cout<<endl;

  prev_permutation(A, A+N);
  cout<<"prev_permutation 后：";
  copy(A, A+N, ostream_iterator<int>(cout, " "));//2 3 4 5 1 6
  cout<<endl;

  next_permutation(A, A+N);
  cout<<"next_permutation 后：";
  copy(A, A+N, ostream_iterator<int>(cout, " "));//2 3 4 5 6 1
  cout<<endl;
  return 0;
}
```

【例 10.17】 一个整型 6 位数由{0,0,1,1,2,3}组成，编程显示它的全排列。

```cpp
//文件名：e10_17.cpp
#include <iostream>
#include <algorithm>
#include <iterator>
using namespace std;
int main()
{
  int a[]={0,0,1,1,3,2};
  sort(a, a+6);
  if(a[0]==0)
  {
```

```
            int * it=upper_bound(a, a+6 , 0);
            swap(a[0], * it);
    }
    do
    {
            copy(a, a+6, ostream_iterator<int>(cout, "\t"));
            cout<<endl;
    }while(next_permutation(a, a+6));
    return 0;
}
```

(1)对整型数组 a 进行升序排序。(2)保证最高位不为 0 且最小。如果 a[0]＝0,则利用 upper_bound 寻找第一个大于 0 的迭代指针位置 it,再交换 a[0]和 * it 的值,即 swap(a[0], * it)。(3)利用 do-while 循环结合 next_permutation 函数生成有效 6 位数的全排列。

10.10 数值算法

10.10.1 主要函数

- accumulate()：累积求和。
- inner_product()：内积求和。
- partial_sum()：创建新序列,每个元素值代表指定范围内该位置前所有元素之和。
- adjacent_difference()：获得相邻元素差集。

函数原型如下所示。

① accumulate

原型：

```
template<class InIt, class T>
    T accumulate(InItfirst, InIt last, T init);
template<class InIt, class T, class Pred>
    T accumulate(InIt first, InIt last, T init,Pred pr);
```

参数说明：

- InIt：输入迭代器。
- T：值类型。
- Pred：二元函数。

第一个模板函数功能是：结果值 result 初始化为 init,对迭代器 it∈[first,last], result＝result＋ * i,最后返回 result。第二个模板函数功能是：结果值 result 初始化为 init,对迭代器 it∈[first,last], result＝pr(result, * it),最后返回 result。

② inner_product

原型:

template<class InIt1, class InIt2, class T>
  T **inner_product**(InIt1 first1, InIt1 last1,InIt2 first2, T val);
template<class InIt1, class InIt2, class T,class Pred1, class Pred2>
  T **inner_product**(InIt1 first1, InIt1 last1, InIt2 first2, T val, Pred1 pr1, Pred2 pr2);

参数说明:

- InIt1,InIt2:输入迭代器。
- T:值类型。
- Pred1,Pred2:二元函数。

第一个模板函数功能是:结果值 result 初始化为 init,$N \in [0, last1 - first1)$,result = result + *(first1+N) * × *(first2+N),最后返回 result。第二个模板函数功能是:结果值 result 初始化为 init,$N \in [0, last1 - first1)$,result = pr1(result, pr2(*(first1+N), (first2+N))),最后返回 result。

③ partial_sum

原型:

template<class InIt, class OutIt>
  OutIt **partial_sum**(InIt first, InIt last,OutIt result);
template<class InIt, class OutIt, class Pred>
  OutIt **partial_sum**(InIt first, InIt last,OutIt result, Pred pr);

参数说明:

- InIt:输入迭代器。
- OutIt:输出迭代器。
- Pred:二元函数。

第一个模板函数功能是:累加和 sum 初值是 *first, *result = sum。$N \in [1, last - first)$,sum = sum + *(first+N), *(result+N) = sum。第二个模板函数功能是:累加和 sum 初值是 *first, *result = sum。$N \in [1, last - first)$,sum = pr(sum, *(first+N)), *(result+N) = sum。

④ adjacent_difference

原型:

template<class InIt, class OutIt>
  OutIt **adjacent_difference**(InIt first, InIt last,OutIt result);
template<class InIt, class OutIt, class Pred>
  OutIt **adjacent_difference**(InIt first, InIt last,OutIt result, Pred pr);

参数说明:

- InIt：输入迭代器。
- OutIt：输出迭代器。
- Pred：二元函数。

第一个模板函数功能是：$*result = *first$，$N \in [1, last-first)$，$*(result+N) = *(first+N) - *(first+N-1)$。第二个模板函数功能是：$*result = *first$，$N \in [1, last-first)$，$*(result+N) = pr(*(first+N), *(first+N-1))$。

10.10.2 示例分析

【例 10.18】 计算 $1+(1 \times 2)+(1 \times 2 \times 3)+\cdots$ 前 10 项和。

```cpp
//文件名：e10_18.cpp
#include <iostream>
#include <numeric>
#include <functional>
#include <vector>
using namespace std;
int main ()
{
    vector<int>v(10);
    for(int i=0; i<10; i++)
    {
        v[i]=i+1;
    }
    vector<int>mid(10);
    partial_sum(v.begin(), v.end(), mid.begin(), multiplies<int>());
    int sum=accumulate(mid.begin(), mid.end(),0);
    cout<<"1+(1×2)+(1×2×3)+…前10项和是:"<<sum<<endl;
    return 0;
}
```

思路是：先用 partial_sum 保存每个子项 $(1, 1 \times 2, 1 \times 2 \times 3, \cdots)$ 的和，再用 accumulate 计算出每个子项的累加和。也就是说：数值运算中，partial_sum 经常与 accumulate 一起运用，partial_sum 保存子项，accumulate 得出总值。

【例 10.19】 计算 $\left(\dfrac{1}{2}\right)^2 + \left(\dfrac{2}{3}\right)^2 + \left(\dfrac{3}{4}\right)^2 + \cdots$ 前 10 项的和。

```cpp
//文件名：e10_19.cpp
#include <iostream>
#include <numeric>
#include <vector>
using namespace std;
```

```cpp
int main ()
{
    vector<float>v(10);
    for(int i=1; i<=10; i++)                                    //为向量赋值
    {
        v.push_back((float)i/(i+1));
        cout<< (float)i/(i+1)<<"\t";
    }
    float sum=inner_product(v.begin(),v.end(),v.begin(),0.0f);  //向量自身做内积
    cout<<sum<<endl;
    return 0;
}
```

第 11 章 STL 应用

本章主要目的是让同学们掌握容器、算法、迭代器的综合应用,进一步体会 STL 容器的作用,算法的汉语语义。本章研究的主要内容如下所示。
(1) 算法的综合运用;
(2) 在数据结构中的应用;
(3) 在 Visual C++ 中的应用。

11.1 算法的综合运用

11.1.1 在多态中的应用

【例 11.1】 已知画图基类 Shape,定义多态函数 Draw,圆类 Circle、正方形类 Square 都从其派生。创建一个含有 Shape 指针的 vector,指向多个 Shape 子对象。首先画出每个 Square,并按其长度进行升序排序;其次画出每个 Circle,并按其半径进行升序排序。

```
//文件名:e11_1.cpp
#include<iostream>
#include<vector>
#include<algorithm>
using namespace std;

class Shape
{
public:
    virtual void Draw()=0;
    virtual int GetMark()=0;
};
class Circle:public Shape                //圆类定义
{
    static int mark;                     //静态变量,Circle 类标识
    float r;                             //半径
public:
```

```cpp
        float GetR(){return r;}
    public:
        Circle(float r)
        {this->r=r;}
        virtual~Circle(){}
        void Draw()
        {
            cout<<"这是圆,半径是："<<r<<endl;
        }
        int GetMark(){return mark;}
        bool operator<(Circle& c)                //排序比较用到此函数
        {
            return r<c.GetR();
        }
};
int Circle::mark=1;                              //静态变量 mark=1,表明是圆类


class Square:public Shape                        //正方形类
{
        static int mark;                         //静态变量,Square 类标识
        float width;
    public:
        float GetWidth(){return width;}
    public:
        Square(float width)
        {this->width=width;}
        virtual~Square(){}
        void Draw()
        {
            cout<<"这是方,半径是："<<width<<endl;
        }
        int GetMark(){return mark;}
        bool operator<(Square& s)                //排序比较用到此函数
        {
            return width<s.GetWidth();
        }
};
int Square::mark=0;                              //静态变量 mark=0,表明是 Square 类


template<class T>
class ShapeCompare                               //排序比较用到的二元函数
{
    public:
```

```cpp
    bool operator()(T * t1,T * t2)
    {
        return * t1< * t2;
    }
};

int main()
{
    vector<Shape * >vecShape;                    //定义 Shape * 向量容器

    Circle * c1=new Circle(10.0f);   Circle * c2=new Circle(7.0f);
    Shape * s1=new Square(10.0f);    Shape * s2=new Square(7.0f);
    vecShape.push_back(c1);          vecShape.push_back(c2);
                                     //Circle,Square 压入 vecShape
    vecShape.push_back(s1);          vecShape.push_back(s2);

    Circle * c3=new Circle(15.0f);   Circle * c4=new Circle(20.0f);
    Shape * s3=new Square(15.0f);    Shape * s4=new Square(20.0f);
    vecShape.push_back(c3);          vecShape.push_back(c4);
                                     //Circle,Square 继续压入 vecShape
    vecShape.push_back(s3);          vecShape.push_back(s4);

    Circle * c5=new Circle(18.0f);   Circle * c6=new Circle(17.0f);
    Shape * s5=new Square(18.0f);    Shape * s6=new Square(17.0f);
    vecShape.push_back(c5);          vecShape.push_back(c6);
    vecShape.push_back(s5);          vecShape.push_back(s6);

    vector<Circle * >vecCircle;                  //定义 Circle * 向量容器
    vector<Square * >vecSquare;                  //定义 Square * 向量容器
    for(int i=0;i<vecShape.size();i++)           //vecShape 向量中哪些是 Circle 或 Square
    {
        Shape * pShape=vecShape.at(i);
        int mark=pShape->GetMark();              //mark:0,Square 对象;mark:1,Circle 对象
        if(mark==0)
            vecSquare.push_back((Square * )pShape);//是方对象,则压入 vecSquare 向量容器
        if(mark==1)
            vecCircle.push_back((Circle * )pShape);//是圆对象,则压入 vecCircle 向量容器
    }
    sort(vecSquare.begin(),vecSquare.end(),ShapeCompare<Square>());
                                     //方对象向量按边长排序
    sort(vecCircle.begin(),vecCircle.end(),ShapeCompare<Circle>());
```

//圆对象向量按半径排序

```
    for(int i=0; i<vecSquare.size(); i++)    //画方
    {
        Square * ps=vecSquare.at(i);
        ps->Draw();
    }
    for(int i=0; i<vecCircle.size(); i++)    //画圆
    {
        Circle * pc=vecCircle.at(i);
        pc->Draw();
    }

    for(int i=0; i<vecShape.size(); i++)    //new---delete 必须配对
    {
        Shape * pShape=vecShape.at(i);
        delete pShape ;
        pShape=NULL;
    }
    return 0;
}
```

(1) 基础类是 Shape、Circle、Square。ShapeCompare 是二元比较函数，用于排序比较。

(2) main 函数中先定义了 vector<Shape *>vecShape 向量，利用多态性质产生了 12 个指向 Circle 或 Square 对象的 Shape * 指针，并把这些 Shape * 指针压入 vecShape 向量中。然后，定义了圆对象指针向量 vector<Circle *>vecCircle 及方对象指针向量 vector<Square *>vecSquare，并遍历 vecShape 向量，如果是圆对象，则压入 vecCircle；如果是方对象，则压入 vecSquare。之后分别对 vecSquare 及 vecCircle 进行 sort 排序，再分别输出，即可实现题目要求：先画出每个 Square，并按其长度进行升序排序；其次画出每个 Circle，并按其半径进行升序排序。

(3) 很明显，遍历 vector<Shape *>vecShape 向量，确定哪些元素是 Circle 对象，哪些元素是 Square 对象是非常关键的一步，示例中是通过多态函数 GetMark 实现的，返回了 Circle 类或 Square 类中静态变量 mark。当 pShape->GetMark()返回 0 时，表明 Shape * 指向 Square 对象；当 pShape->GetMark()返回 1 时，表明 Shape * 指向 Circle 对象。

(4) 当对 vecCircle、vecSquare 向量 sort 排序时，调用了二元函数类 ShapeCompare，由于它即可对 Circle 对象排序，又可对 Square 对象排序，因此该类一定是一个模板类，事实确实如此。在 bool operator()(T * t1, T * t2)中有对 * t1 < * t2 的判断，因此必须重载基本类 Square、Circle 中的 operator< 函数，Square 按边长比较，Circle 按半径比较。有的同学可能想：把 sort(vecSquare.begin(), vecSquare.end(), ShapeCompare<Square>()) 改为 sort(vecSquare.begin(), vecSquare.end())，即去掉二元函数 ShapeCompare，其余不变，这

样不更简单吗？其实是错误的。因为 vecSquare 是 Square * 元素的容器，sort(vecSquare.begin(),vecSquare.end())仅是对指针值大小排序了，根本没有运行 operator<函数，我们需要的是对指针指向的对象进行排序，因此用 ShapeCompare 二元函数作为过渡，再比较指针指向对象大小是必要的。

（5）注意，new 和 delete 在 STL 中也必须配对，有 new，那么在适当位置就应该有 delete。因此在最后一定要再遍历一遍容器对象指针元素，用 delete 把真实对象的物理空间释放掉。

11.1.2 set、map 应用

【例 11.2】 一篇英文文本文件，要求：创建一个集合，保存文件中出现的单词；创建一个映射，保存单词及该单词在文件中出现的次数。

```cpp
//文件名：e11_2.cpp
#include<iostream>
#include<fstream>
#include<sstream>
#include<string>
#include<set>
#include<map>

using namespace std;

class CWord
{
    string word;
public:
    CWord(string word)
    {
        this->word=word;
    }
    string GetWord()const   {return word;}
    bool operator<(const CWord & w) const            //用于添加集合
    {
        return word<w.GetWord();
    }
    bool operator==(const string & s)const           //用于查询
    {
        return word==s;
    }
```

```cpp
class CWordSet
{
    set<CWord>wordSet;
public:
    bool AddString(string s)                              //单词集合添加
    {
        wordSet.insert(CWord(s));
        return true;
    }
    void Show(ostream & os)                               //单词集合显示
    {

        set<CWord>::iterator it=wordSet.begin();
        int n=0;
        while(it!=wordSet.end())
        {
            os<<(*it).GetWord()<<"\t";
            n++;
            if(n%8==0)
            {
                os<<endl;
                n=0;
            }
            it++;
        }

    }
};

class CWordMap
{
    map<CWord,int>wordMap;                                //单词-----出现次数
public:
    bool AddString(string s)
    {
        map<CWord,int>::iterator it=wordMap.find(s);
        if(it==wordMap.end())
        {
            pair<CWord,int>p(CWord(s),1);
```

```
            wordMap.insert(p);
        }
        else
        {
            (*it).second+=1;
        }

        return true;
    }

    void Show(ostream & os)                          //输出函数
    {
        map<CWord,int>::iterator it=wordMap.begin();
        while(it!=wordMap.end())
        {
            string ss=((*it).first).GetWord();
            int n=(*it).second;
            os<<((*it).first).GetWord()<<"\t"<<(*it).second<<endl;;
            it++;
        }
    }
};

int main()
{
    CWordSet   wordSet;
    CWordMap   wordMap;
    int pos=0;
    string s="";
    string delimset=",.";                            //待拆分字符集合 delimset
    ifstream in("d:\\data.txt");
    while(!in.eof())
    {
        getline(in,s);                               //取一行文本
        if(s=="")                                    //若空,取下一行文本
            continue;

        pos=0;
        while((pos=s.find_first_of(delimset,pos))!=string::npos)
                                                     //把串中等于delimset位置处字符都用空格" "代替
            s.replace(pos,1," ");
```

```
            istringstream stringeam(s);
            while(!stringeam.eof())                          //按空格拆分,获得每个单词
            {
                stringeam>>s;
                if(s=="")                                    //若还有空格,则略去
                    continue;
                wordSet.AddString(s);
                wordMap.AddString(s);
            }
        }
        in.close();

        cout<<"单词集合为:"<<endl;
        wordSet.Show(cout); cout<<endl;
        cout<<"单词及出现次数为:"<<endl;
        wordMap.Show(cout);   cout<<endl;
        return 0;
    }
```

(1) CWord 是基础类。按题目要求需要一个单集合类及单映射类,因此有了 CWordSet 及 CWordMap 类。CWordSet 类 AddString 函数负责向 set<CWord>wordSet 中添加新单词,由于 set 中可自动判断并添加非重复元素,因此直接用 set 中 insert 函数添加即可,但必须重载基础类 CWord 类中 bool operator<(const CWord& w)函数,insert 函数可自动根据该接口函数确定待插入单词是否是集合中的重复元素,从而确定是否添加。CWordMap 类 AddString 函数负责向 map<CWord,int> wordMap 中添加单词及其出现次数。因此要先判定该单词是否已在 wordMap 中,若不在,则添加,并初始化计数为 1;若已存在,则获得该单词映射值计数,加 1 即可。再判断单词是否在 wordMap 中,即代码"map<CWord,int>::iterator it=wordMap.find(s)"用到了基础类 CWord 中重载操作符 bool operator==(const string & s)。

(2) 在主程序中主要理解如何在英文文本文件中提取英文单词,转化成编程模型是如何在字符串中拆分出单词,而字符串中可能有各种标点符号等情况,那么怎么办呢?例如 s="Hello,how are you?"。算法是:首先定义拆分字符串 string delim=",?";其次查询待查字符串中是否包含 delim 中单个字符,若有则用空格代替,s="Hello how are you",则单词之间全用空格分开了,最后完全按空格拆分就可得到每个单词了。

11.1.3 ini 文件解析

【例 11.3】 利用 STL 解析 ini 文件。

ini 文件是技术人员经常用到的一种系统配置方法,如何读取和快速识别 ini 文件中的内容实现起来比较烦琐,而 STL 是一个很好的解决方法,省去了许多底层函数的编制。

ini 文件格式示例如表 11.1 所示。

表 11.1 ini 文件格式示例

```
[mydatabase]
connect=jdbc:odbc:mysrc
user=administrator
pwd=123456

[mytable]
tablename=student
tablefields=studno,studname,age,birthday
```

主要有三种元素：section、key 和 value。key-value 也叫键值对，位于等号左右两侧。左侧叫做 key，即关键码；右侧叫做值，它们是成对出现的。section 是由中括号"[ ]"标识的。一个 ini 文件是由多个 section 组成的，一个 section 是由多个 key-value 键值对组成的。我们的任务就是把 ini 文件中的信息以一定的方式保存在程序结构中，代码如下所示。

```cpp
//文件名：e11_3.cpp
#include<iostream>
#include<fstream>
#include<string>
#include<map>
using namespace std;
class MySection
{
    string section;
    map<string,string>mapKey;                         //一个 section
public:                                                //由多个键值对组成
    MySection(string section)
    {this->section=section;}

    bool AddKeyValue(string key,string value)         //加 key-value 键值对函数
    {
        pair<string,string>p(key,value);
        mapKey.insert(p);
        return true;
    }
    void Show(ostream & os)          //显示 section,做测试用,看是否已保存到相应变量中
    {
        os<<section<<endl;
        map<string,string>::iterator it=mapKey.begin();
        while(it!=mapKey.end())
        {
```

```cpp
            os<<"\t"<<(*it).first<<"="<<(*it).second<<endl;
            it++;
        }
    }
};

class MySectionCollect                          //一个 ini 文件
{
    map<string,MySection>mapSection;            //由多个 section 组成
public:
    bool AddSection(string strSection)          //加一个 section
    {
        pair<string,MySection>p(strSection,MySection(strSection));
        mapSection.insert(p);
        return true;
    }
    MySection * GetSection(string strSection)   //根据 section 串标识获得 section 对象
    {
        map<string,MySection>::iterator it=mapSection.find(strSection);
        return &((*it).second);
    }
    void Show(ostream & os)                     //显示各 section 内容,看是否读取正确
    {
        map<string,MySection>::iterator it=mapSection.begin();
        while(it!=mapSection.end())
        {
            ((*it).second).Show(os);
            it++;
        }
    }
};

class ReadIni
{
    string strPath;                             //读 ini 文件路径
    MySectionCollect&  collect;                 //把 ini 文件内容读入到的对象
public:
    ReadIni(string strPath,MySectionCollect& collect):
            strPath(strPath),collect(collect)
            {}
    void Trim(string& s)                        //是 s 去空格
```

```cpp
    {
        if(s!="")
        {
            s.erase(0,s.find_first_not_of(" "));        //删除左空格
            if(s!="")
                s.erase(s.find_last_not_of(" ")+1);     //删除右空格
        }
    }
    string GetSection(string strText)                   //获得去掉[]后的section串
    {
        strText.erase(0,strText.find_first_not_of("["));  //删除左"["
        strText.erase(strText.find_last_not_of("]")+1);   //删除右"]"
        return strText;
    }
    void GetPair(string strText,string& key,string& value)  //获得键key,值value
    {
        int pos=strText.find("=");
        key=strText.substr(0,pos);
        value=strText.substr(pos+1,strText.length()-pos-1);
        Trim(key);
        Trim(value);
    }
    void Process()
    {
        string strLine="";
        string strSection="";
        string strKey="";
        string strValue="";
        MySection * pSection=NULL;

        ifstream in(strPath.c_str());
        while(!in.eof())
        {
            getline(in,strLine);
            Trim(strLine);
            if(strLine=="") continue;                   //若空格,则略去,返回while头

            if(strLine.at(0)=='[')                      //若是section标识
            {
                strSection=GetSection(strLine);         //去掉[]
                collect.AddSection(strSection);         //加一个section
                pSection=collect.GetSection(strSection);
                                    //获得当前MySection对象,用于添加key-value键值对
```

```cpp
            }
            if(strLine.at(0)!='[')                    //若是 key-value 键值对
            {
                GetPair(strLine,strKey,strValue);    //获得 key、value 值
                pSection->AddKeyValue(strKey,strValue);
                                                      //向当前 section 对象添加 key-value 键值对
            }
        }
        in.close();
    }
};

int main()
{
    string path="d:\\data.ini";

    MySectionCollect collect;                         //定义 ini 文件映射对象
    ReadIni ri(path,collect);                         //读 ini 文件类初始化

    ri.Process();                                     //读 ini 文件过程
    collect.Show(cout);                               //显示信息,看是否读取正确
    return 0;
}
```

（1）从语义上来说，一个 section 由多个 key-value 键值对组成，一个 ini 文件由多个 section 组成。因此前者与基本类 MySection 相对应，后者与 MySection 集合类 MySectionCollect 相对应。而 ReadIni 是读 ini 文件类，主要功能是根据 ini 配置文件填充 MySectionCollect 对象。

（2）ReadIni 类中的 Process 函数是负责读 ini 文件算法的，理解好此算法对理解 MySection、MySectionCollect 类中所定义的函数有很大的帮助。该算法描述如下所示。

```
1. 打开 ini 文件输入流
2. while(输入流非空)
3.      strLine<-读一行字符串
4.      strLine 去左右空格
5.      if strLine 是空格 then continue
6.      if strLine 第 1 个字符是"[",表明是新的 section then
7.          获得新 section 标识串
8.          根据该标识串添加新 section
9.          用变量保持住新添加 section 对象,为加 key-value 键值对做准备
10.     if strLine 第 1 个字符不是"[",表明是新的 key-value 对 then
11.         获得 key 及 value 值
12.         用 9 保持的 section 对象,添加 key-value 键值对对象
13. end while
```

(3) 该示例只是生成了 MySectionCollection 对象用以封装 ini 配置文件的内容,这也是最关键的一步,以此为基础可以扩展好多功能。例如常用的一个功能是根据 section 字符串及 key 键而想获得值 value,只需在 MySection 类中及 MySectionCollect 类中各增加一个函数,如下所示。

```
string MySection::Find(string key)
{
    map<string,string>::iterator it=mapKey.find(key);
    return (* it).second;
}

string MySectionCollect::Find(string section,string key)
{
    MySection * pSection=GetSection(section);
    return pSection->Find(key);
}
```

可以看出用 map 进行查询既简洁,速度又较 vector、list 容器快,因此在 MySection、MySectionCollect 类中有两个主要的 map 成员变量,而不是 vector、list 成员变量。

11.1.4 综合查询

【例 11.4】 已知 4 个文本文件:学生信息 stud.dat,每行信息包括学号、姓名、年龄、性别;课程信息 cour.dat,每行信息包括课号、课名;成绩信息 grade.dat,每行信息包括学号、课号、成绩。编功能类,可以完成下述查询功能。

(1) 根据学号查询,输出学号、姓名、课程名、成绩。
(2) 哪些学生选了所有课程,输出其学号。
(3) 查询每门课程的平均成绩,结果按平均成绩升序排列,平均成绩相同时,按课程号降序排列。
(4) 查询每门课程最好的前两名同学的学号。

本示例是针对查询函数的综合运用,所用到的基本类如下所示。

```
//文件名:e11_4.cpp(本示例所有程序都在该文件下)
struct Student                                      //学生信息,针对 stud.dat
{
    string sno;                                     //学号
    string sname;                                   //姓名
    int    sage;                                    //年龄
    string ssex;                                    //性别
    bool operator==(const string & sno) const       //为查询所调用
    {
```

```cpp
        return this->sno==sno;
    }
};
struct Course                              //课程信息,针对 cour.dat
{
    string cno;
    string cname;
    bool operator==(const string & cno) const
    {
        return this->cno==cno;
    }
};
struct SC                                  //成绩信息,针对 grade.dat
{
    string sno;
    string cno;
    int grade;
};
```

查询是围绕上述三个基本对象的集合类展开的,因此查询类中一定有三个集合对象,如下所示。

```cpp
class MyFind
{
    vector<Student>   &vecStud;
    vector<Course>    &vecCour;
    vector<SC>        &vecSC;
    //其他成员变量定义
public:
    MyFind(vector<Student>& s,vector<Course>& c,vector<SC>& sc)
             :vecStud(s),vecCour(c),vecSC(sc)
    {}
    //其他函数定义
};
```

可以看出三个集合对象是引用变量,也就是说,在产生有效查询 MyFind 对象之前,三个真实集合对象已经存在,通过构造函数传入,并可在 MyFind 对象中引用使用。为了突出类中三个引用集合对象,定义的其他成员变量及函数都略去了,在随后的说明中一一论述。

(1) 根据学号查询,输出学号、姓名、课程名、成绩。

若假设各文件信息如表 11.2 所示。

表 11.2 各个文本文件假设内容

学生				成绩			课程	
学号	姓名	年龄	性别	学号	课号	成绩	课号	课名
1001	张	20	男	1001	c001	90	c001	语文
1002	李	21	女	1002	c001	85	c002	数学
1003	王	19	男	1003	c001	80	c003	外语
1004	赵	18	男	1001	c002	89		
1005	周	22	女	1001	c003	86		
1006	孙	23	男	1003	c003	59		

若想查询学号"1001"的姓名、课程名、成绩,姓名应在学生文件中查询得到,课程名应在课程文件中查询获得,成绩应在成绩文件中查询获得,因此这三个文件在查询中都会用到。学生文件和成绩文件通过学号关联,课程文件和成绩文件通过课程号相关联。如果这三个文件是数据库中的表,那么就容易实现了,用 SQL 语句就可以方便完成,其实是 DBMS 完成了具体的工作。那么现在同学们面对的是文件,必须自己完成 DBMS 的工作,尽管编制的功能可能还非常单薄,但对同学们理解将来数据库编程,特别是理解 DBMS 功能、SQL 语句优化等能起到潜移默化的促进作用。

实现该功能源码如下所示。

```
struct ResultView1
{
    string sno;                              //学号
    string sname;                            //姓名
    map<string,int>mapGrade;                 //课程-成绩
};
class MyCompare1
{
    string sno;
public:
    MyCompare1(string sno)
    {this->sno=sno;}
    bool operator()(const SC & s) const
    {
        return sno!=s.sno;
    }
};

//根据学号查询姓名、选修课程名及对应成绩
bool MyFind::GetFactor1Result(string sno,ResultView1& v)
```

```
    {
        vector<Student>::iterator it=find(vecStud.begin(),vecStud.end(),sno);
        if(it==vecStud.end())                                    //若没有此学号,则返回
            return false;
        v.sno=(*it).sno;
        v.sname=(*it).sname;

        vector<SC>vec;
        remove_copy_if(vecSC.begin(),vecSC.end(),back_inserter(vec),MyCompare1(sno));
        for(int i=0; i<vec.size(); i++)
        {
            vector<Course>::iterator it=find(vecCour.begin(),vecCour.end(),vec[i].cno);
            pair<string,int>p((*it).cname,vec[i].grade);
            v.mapGrade.insert(p);
        }
        return true;
    }
```

新封装了一个结果视图结构体 ResultView1,成员是 sno 学号、sname 姓名,由于一个学生可学多门课程,每门课程有一个成绩,因此有一个 map<string,int>课程映射成绩的集合对象 mapGrade。

MyFind 类中增加了一个 GetFactor1Result(string sno,ResultView1 & v)函数,sno 代表学号,v 是 ResultView1 结果视图类的引用。在该函数中主要是根据学生集合,通过学号,利用 find 函数查询出对应姓名;通过成绩集合,利用 remove_copy_if 函数查询出 sno 学号学生学了哪些 cno 课号的课程,把结果存放在临时向量 vec 中。再把临时变量 vec 与课程集合进行关联,利用 find 函数查询到所学某门课程号对应的课程名。这样学号、姓名、课程名、成绩都通过算法得到了,填充 ResultView1 的引用变量 v 即可。

另外还增加了一个一元函数类 MyCompare1,当执行 remove_copy_if 时需要调用它,从而获知 sno 学号学生学了哪些 cno 课程。

(2) 哪些学生选了所有课程,输出其学号。

方法 1:

```
class MyCompare2
{
public:
    bool operator()(const SC& s1,const SC& s2) const
    {
        return s1.sno<s2.sno;
    }
};
//哪些学生选了所有课程,获得其学号
bool MyFind::GetFactor2Result(vector<string>& vResult)
{
```

```
        int nCount=vecCour.size();                    //课程表有多少门课
        sort(vecSC.begin(),vecSC.end(),MyCompare2());  //成绩表按学号升序排序

        int pos=0;
        typedef vector<SC>::iterator FwdIt;
        FwdIt begin=vecSC.begin();                     //获得排序后第一个学生的始末指针
        pair<FwdIt,FwdIt>it;
        while(begin!=vecSC.end())
        {
            it=equal_range(begin,vecSC.end(),vecSC[pos],MyCompare2());
            int nSize=it.second-it.first;
            if(nSize==nCount)                          //表明该学生学了所有课程
            {
                SC * pSC=it.first;
                vResult.push_back(pSC->sno);
            }
            pos+=nSize;                                //查询位置偏移
            begin=it.second;
        }
        return true;
    }
```

基本思想是：先利用课程向量 vecCour 计算有多少门课程，之后利用 sort 函数对成绩向量 vecSC 按学号进行升序排序，其间用到了二元函数 MyCompare2。排序后，学号相同的记录就紧挨在一起了。之后利用 equal_range 函数获得相同学号不同记录的始末位置。根据始末位置就能计算出该学生学几门课，若数量与总课程数量相等，则得到符合条件的一个解。

当然，还可以有其他思路求出哪些学生都选了课程，如方法 2 所示。

方法 2：

```
    bool GetFactor22Result(vector<string>& vResult)
    {
        map<string,int>mymap;                          //定义："学号—课程数"映射
        map<string,int>::iterator it;
        pair<string,int>p;
        for(int i=0; i<vecSC.size(); i++)              //顺序遍历 vecSC 成绩向量
        {
            if(mymap.empty())
            {
                p.first=vecSC[0].sno,p.second=1;
                mymap.insert(p);
            }
            else
```

```
            it=mymap.find(vecSC[i].sno);
            if(it==mymap.end())
            {
                p.first=vecSC[0].sno,p.second=1;
                mymap.insert(p);
            }
            else
            {
                (*it).second+=1;
            }
        }

        int nCount=vecCour.size();              //课程表有多少门课
        it=mymap.begin();
        while(it!=mymap.end())
        {
            if((*it).second==nCount)
                vResult.push_back((*it).first);
            it++;
        }
        return true;
    }
```

基本思想是：定义 map<string,int> mymap，即"学号—课程数"映射。顺序搜索一遍成绩向量 vecSC，对其中的每一个学号 sno，利用 map 中 find 函数查询在"学号—课程数"映射 mymap 中是否存在。若不存在，则利用 map 中函数 insert 插入一条新记录，课程数初始化为 1；若存在，则把课程数加 1 即可。

总之，可思考的问题还很多。本条查询仅是获得选了所有课程学生的学号信息，那么如果还想得到姓名、相应课程名及成绩等，都是同学们值得深思的问题。千万不要为了做题而做题，而要举一反三，融会贯通。

（3）查询每门课程的平均成绩，结果按平均成绩升序排列，平均成绩相同时，按课程号降序排列。

```
struct ResultView3                              //结果显示视图项：课程号+平均成绩
{
    string cno;
    float average;
    //排序调用接口：按成绩升序，若成绩同，按课程号降序
    bool operator<(const ResultView3 & rv) const
    {
```

```cpp
            bool bRet=false;
            if(average<rv.average)
                bRet=true;
            else
            {
                if(average==rv.average)
                {
                    bRet=cno>rv.cno? true:false;
                }
            }
            return bRet;
        }
};
class MyCompare3                                          //成绩向量按课号升序排列接口
{
public:
    bool operator()(const SC & s1,const SC & s2) const
        {
            return s1.cno<s2.cno;
        }
};
class MySum3                                              //accumulate函数调用接口
{
public:
    float operator()(const float & result,const SC & s) const
        {
            return result+s.grade;
        }
};
//查询每门课程的平均成绩,结果按平均成绩升序排列,平均成绩相同时,按课程号降序排列
bool MyFind::GetFactor3Result(vector<ResultView3> & vResult)
{
    sort(vecSC.begin(),vecSC.end(),MyCompare3());         //成绩表按课程号升序排序
    typedef vector<SC>::iterator MyIt;
    pair<MyIt,MyIt>p;
    MyIt begin=vecSC.begin();
    ResultView3 rv3;
    while(begin!=vecSC.end())
    {
        p=equal_range(begin,vecSC.end(),*begin,MyCompare3());
        rv3.cno=(*begin).cno;
        rv3.average=accumulate(p.first,p.second,0.0f,MySum3())/(p.second-p.first);
```

```cpp
        vResult.push_back(rv3);

        begin=p.second;
    }
    sort(vResult.begin(),vResult.end());
    return true;
}
```

基本思想是：先对成绩向量 vecSC 按课程号升序排列，利用的函数是 sort，排序后相同课程号的学生成绩记录就紧挨在一起了；再利用 equal_range 函数，分别确定每门课程的起始结束位置，利用 accumulate 函数求出每门课程的成绩总和，进而求出平均值。

（4）查询每门课程成绩最好前两名同学的学号。

```cpp
struct ResultView4
{
    string cno;
    vector<string>vecSno;
};
class MyCompare4                        //成绩向量按课程号升序排列,若课程号相同,成绩按降序排列
{
public:
    bool operator()(const SC & s1,const SC & s2) const
    {
        bool bRet=false;
        if(s1.cno<s2.cno)
            bRet=true;
        else
        {
            if(s1.cno==s2.cno)
                bRet=s1.grade>s2.grade? true:false;
        }
        return bRet;
    }
};
bool GetFactor4Result(vector<ResultView4>& vResult)
{
    sort(vecSC.begin(),vecSC.end(),MyCompare4());//成绩表按课程号升序,成绩降序排序
    typedef vector<SC>::iterator MyIt;
    pair<MyIt,MyIt>p;
    MyIt begin=vecSC.begin();
```

```
        while(begin!=vecSC.end())
        {
            p=equal_range(begin,vecSC.end(),*begin,MyCompare3());
            ResultView4 rv;
            rv.cno=(*begin).cno;
            rv.vecSno.push_back((*begin).sno);
            begin++;
            rv.vecSno.push_back((*begin).sno);

            vResult.push_back(rv);
            begin=p.second;
        }
        return true;
}
```

基本思想是：先对成绩向量 vecSC 利用 sort 函数按课程号升序，成绩降序排列，再利用 equal_range 函数确定每门课程成绩记录的始末位置。从起始位置开始两个元素可获得成绩最高学生的学号信息，而末位置作为下一门课程学生记录的起始位置。

上述功能实现利用了 sort 全排序，这在数据量大时是很消耗时间的，换个角度思考可发现完成题中要求可不用全排序。先看以下代码，再来分析，如下所示。

```
class MyCompare42                     //按课程号划分接口函数
{
    string cno;
public:
    MyCompare42(string cno)
    {this->cno=cno;}
    bool operator()(const SC& s) const
    {
        return cno==s.cno;
    }
};
class MyCompare43                     //局部排序接口函数
{
public:
    bool operator()(const SC & s1,const SC & s2) const
    {
        return s1.grade>=s2.grade;
    }
};

bool MyFind::GetFactor42Result(vector<ResultView4>& vResult)
{
```

```cpp
        typedef vector<SC>::iterator MyIt;
        MyIt begin=vecSC.begin();

        while(begin!=vecSC.end())
        {
            MyIt mid=partition(begin,vecSC.end(),MyCompare42((*begin).cno));    //先划分
            partial_sort(begin,begin+2,mid,MyCompare43());                      //局部排序
            //nth_element(begin,begin+2,mid,MyCompare43());  //或者用该行代替 partial_sort
            ResultView4 rv;
            rv.cno=(*begin).cno;
            rv.vecSno.push_back((*begin).sno);
            begin++;
            rv.vecSno.push_back((*begin).sno);

            vResult.push_back(rv);
            begin=mid;
        }
        return true;
}
```

该方法基本思想是："先划分,再局部排序"。即先通过 partition 函数根据跟踪的课程号把成绩向量 vecSC 划分成两部分。那么,按顺序来说,第一部分都是某课程号的学生成绩记录,对该部分仍进行 partial_sort 局部排序,而不是 sort 排序即可。由于仅是求成绩最好的两个学生学号,对这两个学生升序、降序都无所谓,因此用 nth_element 函数也可以实现划分功能,完成查询要求,但效率要比 partial_sort 更高。

测试程序仅编制了第一种查询的测试代码,即显示学号为 1001 学生的姓名、课程名、成绩。同学们可据此扩充,把其他的查询功能加进来。测试程序代码如下所示。

```cpp
#include<iostream>
#include<fstream>
#include<sstream>
#include<vector>
#include<string>
#include<map>
#include<algorithm>
#include<numeric>
using namespace std;

int main()
{
    vector<Student>   vecStud;        //定义学生集合
    vector<Course>    vecCour;        //定义课程集合
    vector<SC>        vecSC;          //定义成绩集合
```

```cpp
    ifstream in("d:\\stud.dat");              //打开学生文本文件
    Student s;
    while(!in.eof())
    {
        in>>s.sno>>s.sname>>s.sage>>s.ssex;
        vecStud.push_back(s);                 //形成学生集合具体数据
    }
    in.close();
    ifstream in2("d:\\cour.dat");             //打开课程文本文件
    Course c;
    while(!in2.eof())
    {
        in2>>c.cno>>c.cname;
        vecCour.push_back(c);                 //形成课程集合具体数据
    }
    in2.close();
    ifstream in3("d:\\sc.dat");               //打开成绩文本文件
    SC sc;
    while(!in3.eof())
    {
        in3>>sc.sno>>sc.cno>>sc.grade;
        vecSC.push_back(sc);                  //形成成绩集合具体数据
    }
    in3.close();

    MyFind obj(vecStud,vecCour,vecSC);        //定义查询对象,传入各集合

    ResultView1 rv1;                          //结果视图对象
    obj.GetFactor1Result("1001",rv1);         //查询某学号对应的姓名、课程名、成绩
    cout<<rv1.sno<<"\t"<<rv1.sname<<endl;
    map<string,int>::iterator it=rv1.mapGrade.begin();
    while(it!=rv1.mapGrade.end())
    {
        cout<<"\t"<<(*it).first<<"\t"<<(*it).second<<endl;
        it++;
    }
    return 0;
}
```

11.2 在数据结构中的应用

11.2.1 全排列应用

【例 11.5】 给定一个 n 个整数的集合 $\mathbf{X}=\{x_1,x_2,\cdots,x_n\}$ 和整数 y，找出和等于 y 的 $\mathbf{X}$ 的子集 $\mathbf{Y}$。比如说，如果 $\mathbf{X}=\{10,20,30,40,50,60\}$，$y=60$，则有三种不同长度的解，它们分别是 $\{10,20,30\}$，$\{20,40\}$，$\{60\}$。

分析：$\mathbf{X}$ 有 6 个元素，相对于三个解，可以用布尔向量表示为 $\{1,1,1,0,0,0\}$，$\{0,1,0,1,0,0\}$，$\{0,0,0,0,0,1\}$。值为 1 的部分是解中的元素。换一个说法就是：从 6 个数据中取 1 个，是否等于 60？从 6 个数据中取 2 个，之和是否等于 60？……从 6 个数据中取 6 个，之和是否等于 60？因此转化为从布尔向量 $\{0,0,0,0,0,1\}$ 开始，到 $\{1,1,1,1,1,1\}$ 结束，哪些是符合条件的布尔向量。那么如何变化布尔向量的取值呢？next_permutation 就是一个比较好的选择。得出算法如下：

1. 设置 X,y,定义结果布尔向量 v
2. 通过 x,得出布尔向量长度 n,并初始化
3. for k←0 to n-1
4.   填充结果布尔向量初值 v[0]~v[n-k-2]都为 0
5.   v[n-k-1]~v[n-1]都为 1
6.   do
7.   If v 为解 then 输出布尔向量
8.   while(有并获得下一个向量)
9. end for

程序代码如下所示。

```cpp
//文件名：e11_5.cpp
#include<iostream>
#include<vector>
#include<algorithm>
using namespace std;
bool ValidOrder(vector<int>::iterator start start,vector<int>::iterator end end,
int * s,int total)
{
    int nSize=end-start;
    int sum=0;
    for(int i=0; i<nSize; i++)
    {
        if(*(start+i)==1)
        {
```

```
            sum+=s[i];
        }
    }
    return sum==total;
}
int main()
{
    int X[]={10,20,30,40,50,60};
    int y=60;
    int n=sizeof(X)/sizeof(int);
    vector<int>v(6);
    for(int k=0; k<=n-1; k++)
    {
        if(n-k-2>=0)
            fill(v.begin(),v.begin()+n-k-2,0);
        fill(v.begin()+n-k-1,v.end(),1);

        do
        {
            if(ValidOrder(v.begin(),v.end(),X,y))
            {
                copy(v.begin(),v.end(),ostream_iterator<int>(cout,"\t"));
                cout<<endl;
            }
        }while(next_permutation(v.begin(),v.end()));
    }
    return 0;
}
```

(1) ValidOrder 函数用来判定当前布尔排列是否满足条件：遍历布尔排列,若值为 1,则累加 X 数组对应的真实值。若最终和为 y,则找到符合条件的一个布尔排列。

(2) 生成布尔排列的规则如下：当判断 X 中是否有一个元素等于 y 时,布尔排列需初始化为{0,0,0,0,0,1},结束时为{1,0,0,0,0,0}；当判断 X 中是否有两个元素之和等于 y 时,布尔排列需初始化为{0,0,0,0,1,1},结束时为{1,1,0,0,0,0},依此类推。我们做的工作仅是完成布尔排列的初始化工作,而连续的生成排列由 next_permutation 函数来完成。另外,本示例中还用到了 fill 填充函数。

【例 11.6】 八皇后问题。

可以陈述如下：如何在 8×8 的国际象棋棋盘上安排 8 个皇后,使得没有两个皇后能互相攻击？如果两个皇后处在同一行、同一列或同一条对角线上,则它们能互相攻击。由此可推广到 n 皇后问题：有 n 个皇后和一个 n×n 棋盘,如何摆放能互相不攻击？

由于棋盘有行、列之分,本示例是按列坐标划分的。如图 11.1 所示,一个合适的列坐标排列是{1,5,8,6,3,7,2,4},因此转化为求集合 X={1,2,3,4,5,6,7,8}中哪些排列的向量符合题中要求。满足要求的向量有两点:(1)不能同行同列。由于在向量 X 中没有重复元素,保证了不能同行、同列条件。(2)不能在对角线上。这一点需要函数判定,不难看出两个皇后处在同一条对角线上,当且仅当 abs(X[i]−X[j])=abs(i−j)。由此得算法如下:

1. 设置初始列坐标向量 X
2. do
3.   if X 有效向量 then 输出 X
4. while(有并获得下一个向量)

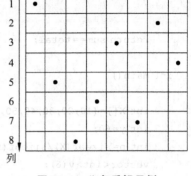

图 11.1 八皇后解示例

程序代码如下所示。

```cpp
//文件名:e11_6.cpp
#include<iostream>
#include<algorithm>
#include<math.h>
using namespace std;
bool ValidOrder(int * start,int * end)
{
    int nSize=end-start;
    for(int i=0; i<nSize-1; i++)
    {
        for(int j=i+1; j<nSize; j++)
        {
            if(j-i==abs(*(start+i)-*(start+j)))
            {
                return false;;
            }
        }
    }
    return true;
}
int main()
{
    int a[]={1,2,3,4,5,6,7,8};
    do
    {
        if(ValidOrder(a,a+8))
        {
```

```
            copy(a,a+8,ostream_iterator<int>(cout,"\t"));
            cout<<endl;
        }
    }while(next_permutation(a,a+8));
    return 0;
}
```

11.2.2 频度问题

【例 11.7】 求整型集合中各元素出现的频度。

算法如下:

1. 设置整型数组 A 中元素,获得元素个数 N
2. sort(A)
3. 初始化查询起始指针,first←A,last←NULL,查询值 find_value←*first
4. do
5.   last←upper_bound(first,A+nSize,find_value)
6.   根据 first,last 查询起始指针显示查询元素及出现频度
7.   first←last
8.   if first!=A+nSize then find_value←*first
9. while first!=A+nSize

程序代码如下所示。

```
//文件名: e11_7.cpp
#include<iostream>
#include<algorithm>
using namespace std;
int main()
{
    int A[]={1,3,2,3,4,3,6,3};
    int nSize=sizeof(A)/sizeof(int);
    sort(A,A+nSize);
    int * first=A;
    int * last=NULL;
    int find_value=A[0];
    do
    {
        last=upper_bound(first,A+nSize,find_value);
        cout<<"find_value="<<find_value<<"\t频度:"<<(float)(last-first)/nSize<<endl;
        first=last;
```

```
        if(first!=A+nSize)
            find_value=*first;
    }while(first!=A+nSize);
    return 0;
}
```

可以看出主要用了 sort,upper_bound 函数非常简洁地解决了问题所需。当然,如果思路变了,所选的函数也不一样。有的同学可能这样想:先 sort 排序,再用 unique_copy 去掉重复值,之后再利用 count 计数,代码如下:

```
int main()
{
    int A[]={1,3,2,3,4,3,6,3};
    int nSize=sizeof(A)/sizeof(int);
    sort(A,A+nSize);
    vector<int>tmp;
    unique_copy(A,A+nSize,back_inserter(tmp));

    for(int i=0; i<tmp.size(); i++)
    {
        int n=count(A,A+nSize,tmp[i]);
        cout<<"find_value="<<tmp[i]<<"\t频度:"<< (float)n/nSize<<endl;
    }
    return 0;
}
```

但是这段代码要比原代码效率低一些,原因如下:(1)它利用 unique_copy 函数获得了待查询的元素集合,而原代码中没有这一步,直接在排序后的集合 A 中边查询边获得待查询的值,主要是"find_value=*first"一行,希望同学们好好理解。(2)它利用了 count 计数,在无序或有序集合中,查找时间是线性关系;而 upper_bound 用在有序中,查找时间是对数关系。既然已经排好序了,那就没有必要用 count 函数了。

因此,STL 虽然提供了强大的函数功能,但是我们必须加以思考,选择效率最高的算法及相应的函数,不能仅为了完成功能而降低了执行效率。再举一个与频度有关的问题:寻找多数元素,令 A[1..n]是一个整数序列,A 中的整数 a 如果在 A 中出现的次数多于 50%,那么 a 称为多数元素。一种解决方法是与本示例相仿,先排序,再计算出各元素出现频度,之后找出最大频度数。另一种方法是既然多数元素指频度大于 50% 的元素,那么第 n/2 大或第 n/2 小元素就可能是多数元素,那么用 nth_element 函数就可以了,根本不必全排序,代码如下:

```
int main()
{
    int A[]={1,3,2,3,4,3,6,3,3};
    int nSize=sizeof(A)/sizeof(int);
```

```
    nth_element(A,A+nSize/2,A+nSize);
    int find_value= * (A+nSize/2);
    int nCount=count(A,A+nSize,find_value);
    float ratio=(float)nCount/nSize;
    if(ratio>0.50f)
    {
        cout<<"多数元素是:"<<find_value<<"\t频度:"<<ratio<<endl;
    }
    else
        cout<<"没有多数元素"<<endl;
    return 0;
}
```

11.2.3 最长公共子序列问题

【例 11.8】 最长公共子序列问题。

即分别给出两个长度为 n 和 m 的字符串 A 和 B,确定在 A 和 B 中最长公共子序列的长度及该串内容。例如 A="zxyxyz",B="xyyzx",那么 xyy 同时是 A 和 B 的长度为 3 的子序列。然而,它不是 A 和 B 最长的公共子序列,因为字符串 xyyz 也是 A 和 B 的公共子序列,长度是 4。由于没有比 4 更长的公共子序列,因此 A 和 B 最长的公共子序列长度是 4。

解决这类问题很好的方法叫做动态规划方法。该方法被广泛用于求解组合最优化问题,使用这种技术的算法不是递归调用自身,而是采取自底向上的方式递推求值,并把中间结果存储起来以便以后用来计算所需要求的解。

为求最长公共子序列,首先寻找一个求最长公共子序列长度的递推公式,令 $A=a_1$,$a_2,\cdots,a_n$ 和 $B=b_1,b_2,\cdots,b_n$,$L[i][j]$ 表示 $a_1,a_2,\cdots,a_n$ 和 $b_1,b_2,\cdots,b_n$ 的最长公共子序列的长度。由于 i=0 或 j=0 时,表明 $a_1,a_2,\cdots,a_n$ 和 $b_1,b_2,\cdots,b_n$ 中的一个或同时可能为空字符串。即如果 i=0 或 j=0,那么 $L[i][j]=0$。由此得如下递推公式。

$$L[i][j] = \begin{cases} 0, & \text{若 } i=0 \text{ 或 } j=0 \\ L[i-1][j-1]+1, & \text{若 } i>0, j>0 \text{ 和 } a[i]=b[j] \\ \max(L[i-1][j],L[i][j-1]), & \text{若 } i>0, j>0 \text{ 和 } a[i] \neq b[j] \end{cases}$$

基于此递推公式,编制程序如下所示。

```
//文件名:e11_8.cpp
#include<iostream>
#include<fstream>
#include<string>
#include<vector>
#include<algorithm>
#include<iterator>
using namespace std;
```

```cpp
        class MyMaxCommonString
        {
            string a;                           //字符串1
            string b;                           //字符串2
            vector<vector<int>>L;               //两字符串二维动态规划表矩阵

        public:
            MyMaxCommonString(string s1,string s2):a(s1),b(s2),
                        L(s1.length()+1,vector<int>(s2.length()+1,0))
            {
            }

            int GetMaxLength()
            {
                int row=L.size();
                int col=L[0].size();
                return L[row-1][col-1];         //L矩阵最后一个元素值是最大公共序列长度值
            }

            string GetMaxCommonString()
            {
                string s="";
                int maxLen=GetMaxLength();
                vector<int>::iterator it;
                if(maxLen>0)                    //若最大公共序列长度值大于0,则一定有公共元素
                {
                    int i=0
                    for(i=0; i<L.size(); i++)
                    {
                        it=find(L[i].begin(),L[i].end(),maxLen);
                        if(it!=L[i].end())
                            break;
                    }
                    int row=i;
                    int col=it-L[i].begin();
                    s+=a.substr(row-1,1);

                    int start=maxLen-1;
                    int end   =start;
                    while(start>=1)
                    {
                        row--,col--;
                        end=L[row][col];
                        if(end!=start)
                        {
```

```cpp
            s+=a.substr(row,1);
            start--;
        }
    }
    reverse(s.begin(),s.end());
}

return s;
}
void Process()                          //按递推公式生成动态变化表
{
    for(int i=1; i<L.size(); i++)
    {
        for(int j=1; j<L[0].size(); j++)
        {
            if(a.at(i-1)==b.at(j-1))
            {
                L[i][j]=L[i-1][j-1]+1;
            }
            else
            {
                L[i][j]=max(L[i][j-1],L[i-1][j]);
            }
        }
    }
}
void Show(ostream & os)                 //输出动态规划表
{
    for(int i=0; i<L.size(); i++)
    {
        copy(L[i].begin(),L[i].end(),ostream_iterator<int>(os,"\t"));
        os<<endl;
    }
}
};
int main()
{
    string s1="xyxxzxyzxy";
    string s2="zxzyyzxxyxxz";
    MyMaxCommonString m(s1,s2);

    cout<<"字符串 1:"<<s1<<endl;
    cout<<"字符串 2:"<<s1<<endl;

    m.Process();
    m.Show(cout);                       //屏幕输出动态规划表
```

```
            int maxlen=m.GetMaxLength();
            string maxstr=m.GetMaxCommonString();
            cout<<"最大公共序列长度:"<<maxlen<<endl;
            cout<<"最大公共序列串:"<<maxstr<<endl;
            return 0;
        }
```

MyMaxCommonString 类中 Process 函数完全是按递推公式编制的,形成了关键矩阵 L,从它可方便分析出最大公共序列的长度及内容。对于示例而言,L 矩阵如表 11.3 所示。

表 11.3 最长公共子序列动态规划递推二维表

	0	1	2	3	4	5	6	7	8	9	10	11	12
0	0	0	0	0	0	0	0	0	0	0	0	0	0
1	0	0	0	1	1	1	1	1	1	1	1	1	1
2	0	0	0	1	1	2	2	2	2	2	2	2	2
3	0	0	1	1	2	2	2	3	3	3	3	3	3
4	0	1	1	2	2	2	3	3	4	4	4	4	4
5	0	1	1	2	2	3	3	4	4	4	4	4	5
6	0	1	2	2	2	3	4	4	4	5	5	5	5
7	0	1	2	2	2	3	4	4	5	5	5	5	5
8	0	1	2	2	3	3	4	4	4	5	5	5	6
9	0	1	2	3	3	3	4	5	5	5	6	6	6
10	0	1	2	3	4	4	4	5	5	6	6	6	6

容易知道,最大公共序列的长度是 L 矩阵的最后一个元素,本示例中值是 6。求最大公共序列串内容是由 GetMaxCommonString 函数完成的。行方向代表 a="zxzyyzxxyxxz",列方向代表 b="xyxxzxyzxy"。a 分别与 b 中第 0、1、2、…直至最后一个字符比较,完成了 L 矩阵的一行行填充。由第 1 行开始递增搜索第 1 次出现 6 的位置,本示例是 row=8,col=12,那么公共序列字符位置一定在表 11.3 中的对角线上。从 row=8,col=12 递减搜索,找出第 1 次出现 5 的行位置,第 1 次出现 4 的行位置,……,第 1 次出现 1 的行位置。因此可得在对角线上关键行位置点{8,6,4,3,2,1},表明对列方向代表字符串 b(索引由 1 开始)而言,最大公共序列字符在 b 上的第 8、6、4、3、2、1 位置上,也就容易求出最大公共序列串的内容了。

11.2.4 大整型数加法、乘法类

【例 11.9】 编一个关于大整型无符号数加法、乘法的类。

整型数一般是 4 个字节,但是有时经常会遇到超大数的运算,那么系统提供的 int、long

就不够用了。因此必须自定义大整型类,设为 MyBigNum,代码如下所示。

```cpp
//文件名：e11_9.cpp
#include<iostream>
#include<string>
#include<deque>
#include<functional>
#include<algorithm>
#include<iterator>
using namespace std;

class MyBigNum
{
    deque<int>v;                                    //用 deque 容器表示大整数
public:
    MyBigNum(){}
    MyBigNum(string strNum)                         //通过字符串建立大整数
    {
        copy(strNum.begin(),strNum.end(),back_inserter(v));
        transform(v.begin(),v.end(),v.begin(),bind2nd(minus<int>(),'0'));
                                                    //字符'1'变成整数 1 要减去'0'
    }
    deque<int>::iterator begin()                    //迭代始指针
    {
        return v.begin();
    }
    deque<int>::iterator end()                      //迭代止指针
    {
        return v.end();
    }
    int size()                                      //容器大小
    {
        return v.size();
    }
    back_insert_iterator<deque<int>>Back_Inserter()
    {
        return back_inserter(v);
    }
    void push_front(int n)                          //前插入元素
    {
        v.push_front(n);
    }
    void push_back(int n)                           //后插入元素
```

```cpp
    {
        v.push_back(n);
    }
    void adjust()                                    //调整使容器每位整型元素值都小于10
    {
        int nSize=v.size();

        for(int i=nSize-1; i>=1; i--)
        {
            int value=v[i];
            if(value<10)
                continue;
            v[i]=value%10;
            v[i-1]+=value/10;
        }

        int value=v[0];                              //处理最高位
        if(value>=10)
        {
            v[0]=value%10;
            value=value/10;
            while(value>0)
            {
                v.push_front(value%10);
                value /=10;
            }
        }
        nSize=v.size();
    }
    MyBigNum Add(MyBigNum & m)
    {
        MyBigNum result;
        int n=size()-m.size();
        if(n>=0)                                     //若大于等于加数位数
        {
            transform(begin()+n,end(),m.begin(),result.Back_Inserter(),plus<int>());
            for(int i=n-1; i>=0; i--)
            {
                result.push_front(*(begin()+i));
            }
        }
        else                                         //若小于加数位数
        {
            transform(begin(),end(),m.begin()-n,result.Back_Inserter(),plus<int>());
```

第 11 章 STL 应用

```cpp
            for(int i=-n-1; i>=0; i--)
            {
                result.push_front(* (m.begin()+i));
            }
            result.adjust();                    //结果调整
            return result;
        }
        MyBigNum Multiply(MyBigNum& m)
        {
            MyBigNum result("0");
            MyBigNum mid;
            for(int i=0; i<m.size(); i++)
            {
                mid= * this;
                for(int j=0; j<i; j++)          //加 0 相当于扩大 10 倍
                {
                    mid.push_back(0);
                }                               //被乘数分别乘以每位乘数
                transform(mid.begin(),mid.end(),mid.begin(),bind2nd(multiplies<int>(),
    * (m.begin()+i)));
                result=mid.Add(result);         //分项之和累加
            }
            return result;
        }
    };
    int main()
    {
        MyBigNum m1("1234567890");
        MyBigNum m2("99999999998");
        MyBigNum result=m1.Add(m2);
        cout<<"1234567890+99999999998=";
        copy(result.begin(),result.end(),ostream_iterator<int>(cout));
        cout<<endl;

        MyBigNum m3("99");
        MyBigNum m4("99999");
        MyBigNum m5=m3.Multiply(m4);
        cout<<"99 * 99999=";
        copy(m5.begin(),m5.end(),ostream_iterator<int>(cout));
        cout<<endl;
        return 0;
    }
```

(1) 如何建立及描述大整型数。一个方案是通过字符串输入建立,通过解析字符串,把每位存入 STL 提供的容器中。那么选择哪个容器呢? vector 还是 deque? 由于一定会遇到进位问题,即对容器头进行进位插入操作,从这一点来说选择 deque<int>双端队列更好,因此 MyBigNum 容器中封装了 deque<int> v 成员变量。从这一点来说,MyBigNum 不能算一个真正的容器,与 stack、queue 相仿,只起到容器适配器的作用。

(2) 关于大整型数加法算法,即类中的 Add 函数。

```
1. 定义结果大整型数变量 result
2. 求被加数位数与加数位数差 n
3. if n>=0 then
4.        从高位开始加数所有位与被加数对应位相加,每位结果和存入 result 后面
5.        把被加数多余位存入 result 前面
6. else
7.        从高位开始被加数所有位与加数对应位相加,每位结果和存入 result 后面
8.        把加数多余位存入 result 前面
9. endif
10. result←result.adjust(),进行位数值调整,保证 result 容器每位值都小于 10
11. 返回 result 给调用者
```

类中 adjust 函数进行大整型数位值调整,保证每位必须小于 10。用表意形式说明如下:设 MyBigNum1={9,9},MyBigNum2={8,8},经过上述 Add 函数 1~9 步结果是 MyBigNum result={17,17},两位都大于 10,因此必须调用 adjust 调整函数,使最终结果为 result={1,8,7}。adjust 算法如下:

```
1. 获得待调整大整型数容器长度 nSize
2. for 从最低位开始到次高位,循环依次调整每一位
3.      若当前位小于 10,转到 2
4.      当前位值等于当前值对 10 取余
5.      利用 while 循环把进位值依次加到前面位的值当中
6. end for
7. if 最高位值大于 10
8.      最高位值等于当前值对 10 取余
9.      改变容器大小,利用 while 循环把进位值依次填充新加位的值
10. endif
```

(3) 关于大整型数乘法算法,即类中的 Multiply 函数。用表意形式说明如下:若 MyBigNum1={1,2,3,4},MyBigNum2={5,6,7},MyBigNum3=MyBigNum1 * MyBigNum2={1,2,3,4}*7+{1,2,3,4,0}*6+{1,2,3,4,0,0}*5。即被乘数后面补充适当个 0 后与乘数每位相乘后再相加。算法如下:

```
1. 定义结果大整型数 result,初始值为 0,用于累加
2. 定义临时大整型数变量 mid
3. for i=0,i<乘数位数,i++
```

4. mid←被乘数大整型数
5. mid 尾部添加适当个数 0 元素
6. mid 中每位值等于 mid 中每位当前值乘以乘数从低位开始的第 i 位数
7. result←mid.Add(result)
8. end for
9. 返回 result 给调用者

11.2.5 矩阵问题

【例 11.10】 编一个关于矩阵加法、乘法的类。

```cpp
//文件名:e11_10.cpp
#include<iostream>
#include<string>
#include<sstream>
#include<vector>
#include<functional>
#include<algorithm>
#include<iterator>
using namespace std;

template<class T>
class MyMatrix
{
public:
    vector<vector<T>>matrix;              //二维矩阵向量
public:
    MyMatrix(int m,int n):matrix(m,vector<T>(n))
    {
    }
    MyMatrix(string s,int m,int n):matrix(m,vector<T>(n))
    {
        istringstream out(s);             //解析字符串至矩阵向量
        char ch;                          //形如"1 2 3,4 5 6,7 8 9"
        for(int i=0; i<m; i++)
        {
            for(int j=0; j<n; j++)
            {
                out>>matrix[i][j];
            }
            out>>ch;
        }
    }
```

```cpp
    int GetMatrixRows()                                //获得矩阵行
    {
        return matrix.size();
    }
    int GetMatrixCols()                                //获得矩阵列
    {
        return matrix[0].size();
    }
    MyMatrix<T>Add(MyMatrix<T>& m)                     //矩阵相加
    {
        int row=GetMatrixRows();
        int col=GetMatrixCols();
        int row2=m.GetMatrixRows();
        int col2=m.GetMatrixCols();
        if(row!=row2||col!=col2)                       //判断是否满足相加条件
            return MyMatrix<T>(0,0);

        MyMatrix<T>tmp(row,col);
        for(int i=0; i<row; i++)
        {
            transform(matrix[i].begin(),matrix[i].end(),m.matrix[i].begin(),tmp.matrix[i].begin(),plus<T>());
        }
        return tmp;
    }
    MyMatrix<T>Minus(MyMatrix<T>& m)
    {
        int row=GetMatrixRows();
        int col=GetMatrixCols();
        int row2=m.GetMatrixRows();
        int col2=m.GetMatrixCols();
        if(row!=row2||col!=col2)                       //判断是否满足相加条件
            return MyMatrix<T>(0,0);

        MyMatrix<T>tmp(row,col);
        for(int i=0; i<row; i++)
        {
            transform(matrix[i].begin(),matrix[i].end(),m.matrix[i].begin(),tmp.matrix[i].begin(),minus<T>());
        }
        return tmp;
    }
    MyMatrix<T>Multiply(MyMatrix<T>& m)
```

```cpp
        {
            int row=GetMatrixRows();
            int col=GetMatrixCols();
            int row2=m.GetMatrixRows();
            int col2=m.GetMatrixCols();
            if(col!=row2)
            {
                return MyMatrix<T>(0,0);
            }
            MyMatrix<T>tmp(row,col2);
            for(int i=0; i<row;i++)
            {
                for(int j=0; j<col2; j++)
                {
                    tmp.matrix[i][j]=0;
                    for(int k=0; k<col; k++)
                        tmp.matrix[i][j]+=matrix[i][k] * m.matrix[k][j];
                }
            }
            return tmp;
        }
        bool Success()                          //是否计算成功,若矩阵大小不为 0,成功
        {
            return(matrix.size()==0? false:true);
        }
        void Show(ostream & os)                 //输出矩阵向量,可向屏幕,可向文件等
        {
            int m=matrix.size();
            int n=matrix[0].size();
            for(int i=0; i<m; i++)
            {
                copy(matrix[i].begin(),matrix[i].end(),ostream_iterator<T>(os,"\t"));
                cout<<endl;
            }
        }
};
int main()
{
    string s="1 2 3,4 5 6,7 8 9";
    string s2="1 2 3 4 5 6 7,5 6 7 8 9 10 11,9 10 11 12 13 14 15";
    MyMatrix<int>m1(s,3,3);
    MyMatrix<int>m2(s2,3,7);
    cout<<"矩阵 m1:"<<endl;
```

```
    m1.Show(cout);
    cout<<"矩阵 m2:"<<endl;
    m2.Show(cout);
    MyMatrix<int>m3=m1.Multiply(m2);
    cout<<"m1 * m2:"<<endl;
    if(m3.Success())
    {
        m3.Show(cout);
    }
    else
    {
        cout<<"不满足矩阵相乘条件"<<endl;
    }
    return 0;
}
```

（1）通过本例学习了如何用 vector 向量描述二维矩阵，若 vector<T>是一维数组，则 vector<vector<T>>就可描述二维矩阵。最好在构造函数中完成矩阵初始化工作，如"MyMatrix(int m,int n):matrix(m,vector<T>(n)){}"。含义是向量 matrix 有 m 行、n 列，之后就可用 matrix[ ][ ]随机访问向量表示的矩阵元素了。

（2）本示例封装了矩阵的加法、减法、乘法运算，并有条件判定两个矩阵是否能进行上述运算，若不能运算，则返回一个大小为 0 的二维矩阵，通过封装的 Success 函数可判定计算成功与否。

11.2.6　回溯问题

【**例 11.11**】　迷宫问题。

求迷宫中从入口到出口的路径是一个经典的程序设计问题。即从入口出发，顺某一方向向前搜索，若能走通，则继续往前走；否则沿原路退回，换一个方向再继续探索，直至所有可能的通路都探索到为止。为了保证在任何位置上都能沿原路退回，显然需要用一个后进先出的栈结构来保存从入口到当前位置的路径。可用图 11.2 加以辅助说明。

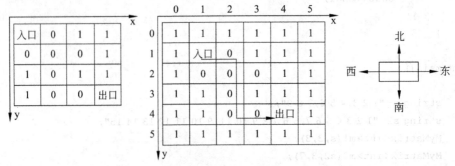

图 11.2　迷宫示意图

图中 0 表示通路,1 表示墙壁;横坐标是 x,纵坐标是 y。从某位置开始只能沿东、南、西、北 4 个方位走,假设它们的标识符是整型数 0,1,2,3,走的优先级是按东南西北依次降低。为了分析问题的方便,特别是编程时边界条件容易处理,在图 11.2 左边图四周再假想加了一层墙壁,就变成了图 11.2 的中间图,入口坐标位置就变成(1,1),出口位置变成(4,4)了。所走的路线由三原组(x,y,方位)组成,从入口到达出口的过程描述如图 11.3 所示。

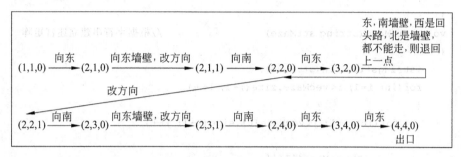

图 11.3　走迷宫过程描述图

很明显,从(3,2,0)位置返回到(2,2,0)位置,需要保持(2,2,0)状态,且具有后进先出的特点,即具有回馈的功能,符合栈特点。为了防止重走原路,另外设置一个标志矩阵,所有元素初始化都为 0,一旦行进到某个位置,则将对应位置置为 1,下次这个位置就不能再走了。迷宫算法程序代码如下所示。

```cpp
//文件名:e11_11.cpp
#include<iostream>
#include<vector>
#include<stack>
#include<string>
#include<sstream>
#include<iterator>
using namespace std;

struct PathUnit                                  //三元组结构
{
    int x;                                       //坐标 x
    int y;                                       //坐标 y
    int direct;                                  //方位特征值。0 东,1 南,2 西,3 北
};

class MyMaze
{
    vector<vector<int>>vecMaze;                  //迷宫描述矩阵
    vector<vector<int>>vecMark;                  //迷宫标志矩阵
public:
    MyMaze(int m,int n):vecMaze(m+2,vector<int>(n+2)),
```

```cpp
        vecMark(m+2,vector<int>(n+2))
        for(int i=0; i<m+2; i++)
        {
            fill(vecMaze[i].begin(),vecMaze[i].end(),1);
        }
    }
    void InitMaze(string strMaze)                    //根据字符串建立迷宫矩阵
    {
        istringstream in(strMaze);
        for(int i=1; i<vecMaze.size()-1; i++)
        {
            for(int j=1; j<vecMaze[0].size()-1; j++)
            {
                in>>vecMaze[i][j];
            }
        }
    }
    void ShowMaze()                                  //显示迷宫矩阵
    {
        for(int i=0; i<vecMaze.size(); i++)
        {
            copy(vecMaze[i].begin(),vecMaze[i].end(),ostream_iterator<int>(cout,"\t"));
            cout<<endl;
        }
    }
    void GetNextPath(const PathUnit & cur,PathUnit & next)  //根据当前位置走向,获得下一结点
    {
        next=cur,next.direct=0;
        switch(cur.direct)
        {
        case 0:
            next.x+=1; break;
        case 1:
            next.y+=1; break;
        case 2:
            next.x-=1; break;
        case 3:
            next.y-=1; break;
        }
    }
    bool GetOuter(const PathUnit & u)                //当前结点是否是出口结点
```

```cpp
    {
        int out_x=vecMaze[0].size()-2;
        int out_y=vecMaze.size()-2;

        bool bRet=((u.x==out_x && u.y==out_y)? true:false);
        return bRet;
    }
    void GetPath()                                      //迷宫算法函数
    {
        stack<PathUnit>st;
        PathUnit unit;
        unit.x=1;
        unit.y=1;
        unit.direct=0;

        st.push(unit);
        bool bOver=false;
        while(!st.empty())
        {
            PathUnit cur=st.top();                      //获得栈顶结点3元组,并出栈
            st.pop();
            PathUnit next;
            while(cur.direct<=3)                        //若有方位可走
            {
                GetNextPath(cur,next);                  //获得下一结点3元组
                if(GetOuter(next))                      //若是出口结点
                {
                    st.push(cur);                       //当前结点入栈
                    st.push(next);                      //下一结点入栈
                    bOver=true;                         //置结束标志
                    break;                              //退出循环
                }
                if(!vecMaze[next.y][next.x] &&!vecMark[next.y][next.x])
                                                        //合法移动且从未走过
                {
                    vecMark[next.y][next.x]=1;          //标志矩阵置1,下次不能再走了
                    st.push(cur);                       //当前结点入栈
                    cur=next;                           //当前结点等于下一结点,从下一结点继续跟踪
                }
                else                                    //换当前结点方位,继续跟踪
                {
                    cur.direct++;
                }
            }                                           //while(cur.direct<=8)
```

```
            if(bOver)break;
        }
        if(bOver)
        {
            while(!st.empty())
            {
                PathUnit & u=st.top();
                cout<<"("<<u.x<<","<<u.y<<","<<u.direct<<")"<<endl;
                st.pop();
            }
        }
        else
        {
            cout<<"there is no path"<<endl;
        }
    }
};

int main()
{
    MyMaze m(6,6);
    string s="";
    s=s+"001011 "+"010011 "+"000111 "+"101111 "+"100001 "+"111100";
                                                    //每个子串后面有空格
    m.InitMaze(s);
    cout<<"显示迷宫矩阵:"<<endl;
    m.ShowMaze();
    cout<<"显示 3 元组代表结点路径:"<<endl;
    m.GetPath();
    return 0;
}
```

11.2.7 字符串表达式

【例 11.12】 编一个功能类,实现字符串表达式求值。

表达式求值是程序设计语言中很重要的一个问题,它的实现是栈应用的典型例子,即"算符优先算法"。算术四则运算的规则是:①先乘除,后加减;②从左算到右;③先括号内,后括号外。

任何一个表达式都是由操作数、运算符和界限符组成的。本示例中为了简便起见,操作数采用浮点常数,运算符限于＋、－、*、/,界限符限于左右括号。

由于该算法中涉及许多知识都很精巧,为了更好说明,把头文件与源文件分开加以描述。代码如下所示。

```cpp
//e11_12.cpp(头文件及源文件内容都在该文件内)
#include<iostream>
#include<sstream>
#include<string>
#include<stack>
#include<vector>
using namespace std;
class MyExpress
{
    string m_strChars;                                    //算符字符串
    vector<vector<int>> m_vComp;                          //优先级比较矩阵
public:
    MyExpress(string m_strChars,string strComps);
    bool GetUnit(string strExpress,int & pos,string & dest,bool& bOper);
                                                          //返回每个操作符或操作数
    char CompareProc(char chStack,char ch);               //优先级比较函数
    float Calc(float value1,float value2,char oper);      //操作数运算函数
    float Process(string strExpress);                     //表达式处理函数
};
```

源程序各个函数说明如下。
(1) 构造函数。

```cpp
MyExpress::MyExpress(string m_strChars,string strComps):
    m_vComp(m_strChars.length()-1,vector<int>(m_strChars.length()))
{
    this->m_strChars=m_strChars;
    //根据字符串创建比较矩阵
    istringstream in(strComps);
    for(int i=0; i<m_vComp.size()-1; i++)
    {
        for(int j=0; j<m_strChars.length(); j++)
        {
            in>>m_vComp[i][j];
        }
    }
}
```

由于操作数在示例中是实数,因此表达式功能本质上是由运算符和界限符组成的,把运算符和界限符组成的集合叫做算符。算符内容及算符间优先级内容描述是非常关键的内

容，因此 MyExpress 类中定义了与之对应的两个成员变量，用字符串 m_strChars 封装算符内容，二维向量 m_vComp 描述算符间运算优先级，MyExpress 构造函数主要是完成这两个变量的初始化。在本示例中，m_strChars 等于串"＋ － ＊ / ( )"，表明 MyExpress 类中可实现有（无）括号的操作数加、减、乘、除运算。那么算符优先级 m_vComp 填充什么内容呢？其实就是把图 11.4(a)内容映射到图 11.4(b)内容，m_vComp 填充的是图 11.4(b)的内容，即任意两个相继出现的算符 b1 和 b2 之间的关系。

b1\b2	+	-	*	/	(	)
+	>	>	<	<	<	>
-	>	>	<	<	<	>
*	>	>	>	>	<	>
/	>	>	>	>	<	>
(	<	<	<	<	<	=

b1\b2	+	-	*	/	(	)
+	1	1	0	0	0	1
-	1	1	0	0	0	1
*	1	1	1	1	0	1
/	1	1	1	1	0	1
(	0	0	0	0	0	0

(a) 逻辑符号表示算符间关系　　　　　(b) 整型数表示算符间关系

图 11.4　算符间优先关系

图 11.4(b)所示内容是通过在构造函数中传入字符串 strComps 并解析完成的，形式如下所示。

```
strComps="110001"+"110001"+      //每个子串后有空格
        "111101"+"111101"+
        "000000";                //最后子串没有空格
```

可以看出 b1 操作符比 b2 操作符少一个右括号，这是因为在本示例算法中没有用到的缘故，下面还有论述。

(2) 字符串表达式处理主函数 Process。

主要定义了两个工作栈，一个是操作数栈，一个是算符栈。关键思想是：当前读取算符与算符栈顶算符进行优先级比较，若当前算符优先级大，则把该运算符压入算符栈；若当前算符优先级小，则在算符栈中弹出该算符，在操作数栈中连续弹出两个操作数，运算结束后把结果再压入操作数栈，具体算法如下所示。

1. 已知表达式字符串 strExpress
2. 初始化操作数栈，算符栈
3. while(true)
4. 　　若已到字符串尾，则退出 while 循环
5. 　　根据算符集搜索 strExpress，获得有效字符串 dest
6. 　　若 dest 是操作数，则压入操作数栈
7. 　　若 dest 是算符，与算符栈顶算符比较
8. 　　若 dest 优先级大，则 dest 入算符栈
9. 　　if dest 优先级小，则
10. 　　　　算符栈弹出算符 ch_oper

11.   操作数栈连续弹出操作数 value1,value2
12.   value1 与 value2 进行 ch_oper 操作,结果入操作数栈
13.   若 ch_oper 不是右括号")",则 ch_oper 入算符栈
14.   否则,逆向运算直到遇见栈顶元素为"("为止,并弹出"("
15.   endif
16. end while
17. 返回操作数栈顶元素,为表达式计算结果

函数源码如下所示。

```cpp
float MyExpress::Process(string strExpress)   //表达式处理函数
{
        string str="("+strExpress+")";
        strExpress=str;

        stack<float> opnum;                //操作数栈
        stack<char> optr;                  //操作符栈

        int pos=0;
        string dest="";
        bool bOper=false;
        while(true)
        {
            bool bRet=GetUnit(strExpress,pos,dest,bOper);
            if(!bRet)                      //若操作数或操作符单元全部取完
                break;

            if(!bOper)                     //若是操作数直接进栈
            {
                float value;
                istringstream is(dest);
                is>>value;
                opnum.push(value);
            }
            else                           //若是操作符
            {
                char ch_oper=dest.at(0);
                char result;
                if(optr.empty())
                    result='<';
                else
                    result=CompareProc(optr.top(),ch_oper);
                switch(result)
                {
```

```cpp
                    case '<':
                        optr.push(ch_oper);
                        break;
                    case '>':
                        char ch_oper2=optr.top(); optr.pop();
                        float value1=opnum.top(); opnum.pop();
                        float value2=opnum.top(); opnum.pop();
                        float value=Calc(value1,value2,ch_oper2);

                        opnum.push(value);
                        if(ch_oper!=')')        //若操作符不是")"
                            optr.push(ch_oper);
                        else                    //若是")",则逆向计算,直到遇见"("为止
                        {
                            ch_oper2=optr.top();
                            while(ch_oper2!='(')
                            {
                                float value1=opnum.top(); opnum.pop();
                                float value2=opnum.top(); opnum.pop();
                                float value=Calc(value1,value2,ch_oper2);

                                opnum.push(value);
                                optr.pop();
                                ch_oper2=optr.top();
                            }
                            optr.pop();
                        }
                        break;
                }
            }
        }                                       //while(true)
        return opnum.top();
    }
```

黑体部分代码是同学们需要多思考的地方。从中可以看出,右括号")"根本就不进入算符栈,图 11.4 中 b1 代表可能进入算符栈的元素,也就可以明白为什么没有右括号")"了。

（3）获得操作数串或操作符串函数 GetUnit。

```cpp
bool MyExpress::GetUnit(string strExpress,int& pos,string& dest,bool& bOper)
                                        //返回每个操作符或操作数字符串
{
    if(pos>=strExpress.length())        //若搜索位置大于等于串长度,返回 false
        return false;
```

```
        bOper=false;                              //默认是操作数标识
        dest=strExpress.substr(pos,1);
        if(m_strChars.find(dest)!=string::npos)   //若是操作符
        {
            bOper=true;                           //置操作符标识为真
            pos+=1;                               //搜索下一个操作数或操作符起始位置
        }
        else                                      //获取操作数
        {
            int end=strExpress.find_first_of(m_strChars,pos+1);
            if(end!=string::npos)
            {
                dest=strExpress.substr(pos,end-pos);
                pos=end;
            }
            else
            {
                dest=strExpress.substr(pos,strExpress.length()-pos);
                pos=strExpress.length();
            }
        }
        return true;
    }
```

该算法很简单,用一个实例更能说明问题。例如若 strExpress="12.5+3",从字符串第 pos=0 位开始,先获得第 0 位字符"1",看它是否在算符集 m_strChars="+ - * /()"中,很明显"1"不在该字符集中,那么以"1"开头的一定是操作数,操作数到哪结束呢？很明显当遇到算符时就结束了,因此只要从 0 位开始查询到第 1 个属于算符的位置,就得到操作数结束位置 end=4,那么截取 strExpress 串[0,4)位置字符,就得到操作数串"12.5"。下一次 pos 从 4 开始搜索,第 4 位字符是"+",属于算符集中字符,因此"+"一定是运算符,以此类推。把这段说明看清楚,函数中的具体代码也就不难理解了。

(4) 优先级比较函数 CompareProc。

```
char MyExpress::CompareProc(char chStack,char ch)   //优先级比较函数
{                                                   //chStack:算符栈顶运算符,ch:当前获得单元运算符
    if(chStack=='#')return'<';
    int pos1=m_strChars.find(chStack);
    int pos2=m_strChars.find(ch);
    return(m_vComp[pos1][pos2]==1?'>':'<');
}
```

(5) 操作数运算函数 Calc。

```
float MyExpress::Calc(float value1,float value2,char oper)   //操作数运算函数
```

```
    {
        float value=0.0f;
        switch(oper)
        {
        case '+':
            value=value1+value2; break;
        case '-':
            value=value1-value2; break;
        case '*':
            value=value1*value2; break;
        case '/':
            value=value1/value2; break;
        }
        return value;
    }
```

(6) 测试函数 main。

```
int main()
{
    string m_strChars="+-*/()";
    string strComps="";
    strComps=strComps+"1 1 0 0 0 1 "+
                      "1 1 0 0 0 1 "+
                      "1 1 1 1 0 1 "+
                      "1 1 1 1 0 1 "+
                      "0 0 0 0 0 0";
    MyExpress obj(m_strChars,strComps);
    string str="(((1+2*3)+4*5)*6)";
    float value=obj.Process(str);
    cout<<"表达式"<<str<<"="<<value<<endl;
    return 0;
}
```

11.2.8 图

【例 11.13】 编一个无向连通图的功能类。

图是一种重要的数据结构，运用综合知识也比较多。本示例主要以如何应用 STL 编制图的深度优先搜索、广度优先搜索、最小生成树加以说明的。对于较短的代码，直接写在了头文件中。对于较长或者着重说明的写在了后续论述中。

```
//e11_13.cpp(头文件及源文件内容都在该文件内)
class MyNode                                              //结点类
```

```cpp
{
    int v;                                  //结点号,关键字,不重复
    string strDiscript;                     //结点描述
public:
    MyNode(int v,string strDiscript);       //构造函数
    {
        this->v=v,this->strDiscript=strDiscript;
    }
    string GetNodeDiscript();               //获得结点描述
    {return strDiscript;}
};

template<class DistType>                    //DistType:权重类型
class MyEdge                                //边类
{
    int head,tail;                          //head:边起始结点号,tail:边结束结点号
    DistType weight;                        //权重
public:
    MyEdge(int head,int tail,DistType weight)
    {
        this->head=head,this->tail=tail;
        this->weight=weight;
    }
    int GetHead()const{return head;}
    int GetTail()const{return tail;}
    DistType GetWeight()const{return weight;}
    bool operator==(int node)               //为了stl find 函数起作用
    {return tail==node;}
    bool operator<(const MyEdge<DistType>& m)const       //为了sort排序起作用
    {return weight<m.GetWeight();}
};

template<class DistType>
class NodeCompare                           //一元函数
{
    int value;
public:
    NodeCompare(int value)
    {this->value=value;}
    bool operator()(const MyEdge<DistType>& m)const
    {return m.GetHead()==value;}
```

```cpp
};

template<class DistType>
class MyMinSpanTree                              //最小生成树类
{
    int vertices;                                //表明是多少点的最小生成树
    vector<MyEdge<DistType>> vec_Edge;
public:
    MyMinSpanTree(int n){vertices=n;}
    bool IsCircuit(int from,int to);             //判定回路
    bool Insert(MyEdge<DistType>& e)             //插入最小边函数
    void Show()                                  //显示最小生成树结果
    {
        for(int i=0; i<vec_Edge.size(); i++)
        {
            cout<<vec_Edge[i].GetHead()<<"\t"<<vec_Edge[i].GetTail()<<"\t"<<vec_Edge[i].GetWeight()<<endl;
        }
    }
};

template<class NodeType,class DistType>
class MyGraph                                    //图功能类
{
private:
    vector<NodeType> vec_vertices;               //顶点集合
    vector<vector<DistType>> vec_matrix;         //邻接矩阵
public:
    MyGraph(string strNodes,string strMatrix,int nSize);
    bool GraphEmpty(){return vec_vertices.empty();}
    int NumberOfVertices(){return vec_vertices.size();    }
    DistType GetWeight(int v1,int v2)            //得出 v1->v2 边权值
    {return vec_matrix[v1][v2];}
    int GetFirstNeighbour(int v);                //给出 v 的第 1 个邻接顶点的位置
    int GetNextNeighbour(int v1,int v2);         //给出 v1 的某邻接顶点 v2 的下一个邻接顶点
    void DFS(int start);                         //从某点开始进行深度优先搜索
    void DFS(int v,vector<int>& visit);
    void BFS(int v);                             //广度优先搜索
    void Kruskal(MyMinSpanTree<DistType>& t);    //克鲁斯卡尔最小生成树算法
    void Show()                                  //显示图的邻接矩阵
    {
```

```
        for(int i=0; i<vec_matrix.size(); i++)
        {
            copy(vec_matrix[i].begin(),vec_matrix[i].end(),ostream_iterator
<DistType>(cout,"\t"));
            cout<<endl;
        }
    }
};
```

MyGraph 类是中心类，所有功能都是围绕此类展开的，下面加以描述。

(1) 构造函数。

图是由顶点及边组成的，因此 MyGraph 类中有两个主要成员变量：一个是顶点的集合向量 vector<NodeType> vec_vertices；一个是描述边信息的二维邻接矩阵 vector<vector<DistType>> vec_matrix。NodeType 是模板参数，在本例中指的是 MyNode 基础类，它有两个成员变量，一个是顶点号 int v，一个是顶点描述串 string strDiscript。DistType 是模板参数，说明权值可能是 int、float 或其他。MyGraph 构造函数就是形成 vec_vertices 及 vec_matrix 的具体值。

```
MyGraph::MyGraph(string strNodes,string strMatrix,int nSize)
                    :vec_matrix(nSize,vector<DistType>(nSize))
{
    //create node vector
    int v=-1;
    string strDiscript;
    istringstream innode(strNodes);
    while(!innode.eof())
    {
        innode>>v>>strDiscript;
        MyNode node(v,strDiscript);
        vec_vertices.push_back(node);
    }
    //create edge matrix
    istringstream in(strMatrix);
    for(int i=0; i<nSize; i++)
    {
        for(int j=0; j<nSize; j++)
        {
            in>>vec_matrix[i][j];
        }
    }
}
```

由代码可知，图的顶点信息及矩阵信息分别隐含于字符串 strNodes 及 strMatrix 中，格式如表 11.4 所示。

表 11.4　顶点信息及邻接矩阵字符串输入格式描述

类　　型	格　　式	说　　明
顶点信息字符串	string strNodes=""; strNodes=strNodes+"0 城市 a 1 城市 b 2 城市 c 3";	0号顶点是城市a,以此类推
邻接矩阵字符串	string strEdge=""; strEdge=strEdge+"0 6 1 5 "+ 　　　　　　　　"6 0 5 0 "+ 　　　　　　　　"1 5 0 5 "+ 　　　　　　　　"5 0 5 0 ";	字符串分行写,由加号连接,与邻接矩阵相仿,由于解析字符串是按空格拆分的,因此每一行字符串(最后一行除外)最后要以空格结束,如"0 6 1 5 "正确,而"0 6 1 5"错误
两者关系	可以互相推出信息,根据邻接矩阵第一行"0 6 1 5",结合顶点信息知道城市0与城市1,2,3相通	

　　本例中是通过字符串来建立图对象的,同学们可以通过重载构造函数,以不同的形式创建图对象。

　　(2) 简单函数。

　　包括判断图空函数 GraphEmpty、顶点个数函数 NumberOfVertices、获得边权重函数 GetWeight。这三个函数的函数体在头文件中给出,另外还有两个函数,用以求邻接顶点号,都是通过 STL find 函数实现的,如下所示。

```
int MyGraph::GetFirstNeighbour(int v)           //给出 v 的第 1 个邻接顶点的位置
{
    int nPos=find_if(vec_matrix[v].begin(),vec_matrix[v].end(),bind2nd(greater<DistType>(),0))
        -vec_matrix[v].begin();
    return(nPos==vec_matrix[v].size()?-1:nPos);    //-1 表示没有邻接顶点
}
int MyGraph::GetNextNeighbour(int v1,int v2)
                        //给出 v1 的某邻接顶点 v2 的下一个邻接顶点
{
    int nPos=find_if(vec_matrix[v1].begin()+v2+1,vec_matrix[v1].end(),bind2nd(greater<DistType>(),0))
            -vec_matrix[v1].begin();
    return(nPos==vec_matrix[v1].size()?-1:nPos);   //-1 表示没有邻接顶点
}
```

　　这两个函数是进行图复杂算法的基础,在下面的深度优先搜索、广度优先搜索、最小生成树函数功能实现中都会遇到。

　　(3) 深度优先搜索(Depth First Search,DFS)。

　　DFS 在访问图中某一起始结点 v 后,由 v 出发,访问它的任意邻接顶点 w_1;再从 w_1 出发,访问与 w_1 邻接但还没有访问过的顶点 w_2;然后再从 w_2 出发,进行类似的访问……直至到达所有的邻接顶点都被访问过的顶点 u 为止。接着,退回一步,退到前一次刚访问过的顶点,看是否还有其他没有被访问的邻接顶点。如果有,则访问此顶点,之后再从此顶点出发,进行与前述类似的访问;如果没有,就再退回一步进行搜索。重复上述过程,直到连通图中所有顶点都被访问过为止。

图 11.5(a)给出深度优先搜索的示例。从顶点 A 出发,开始一次深度优先搜索,可以到达连通图的所有顶点。各顶点旁边附加的数字表明了各顶点访问的次序。图 11.5(b)给出在深度优先搜索的过程中访问过的所有顶点和遍历时经过的边。这是由 n−1 条边连结了所有 n 个顶点而形成的图,称此图为原图的优化生成树。下面给出深度优先搜索的递归算法函数。

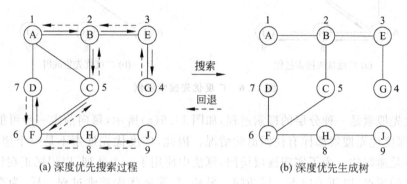

(a) 深度优先搜索过程　　　　　　　　(b) 深度优先生成树

图 11.5　深度优先搜索示例

```
void MyGraph::DFS(int start)                    //从某点开始进行深度优先搜索
{
    vector<int>visit(vec_vertices.size());
    DFS(start,visit);
}
void DFS(int v,vector<int>& visit)
{
    cout<<v<<"\t"<<vec_vertices[v].GetNodeDiscript()<<endl;
    visit[v]=1;                                 //访问标志改为已访问过
    int w=GetFirstNeighbour(v);                 //找顶点 v 的第一个邻接顶点
    while(w!=-1)                                //有邻接顶点
    {
        if(!visit[w])                           //若未访问过,从 w 递归访问
            DFS(w,visit);
        w=GetNextNeighbour(v,w);                //找顶点 v 的下一个邻接顶点
    }
}
```

(4) 广度优先搜索(Breadth First Search,BFS)。

在访问了起始顶点 v 之后,由 v 出发,依次访问 v 的各个未曾被访问过的邻接顶点 w_1,w_2,…,w_n。再从这些访问过的顶点出发,访问它们的所有还未被访问过的邻接顶点……如此做下去,直到图中所有顶点都被访问到为止。图 11.6(a)给出一个从顶点 A 出发进行广度优先搜索的例子。图中各顶点旁边附的数字标明了顶点访问的顺序。图 11.6(b)给出经由广度优先搜索得到的广度优先生成树,它由遍历时访问过的 n 个顶点和遍历时经历的 n−1 条边组成。

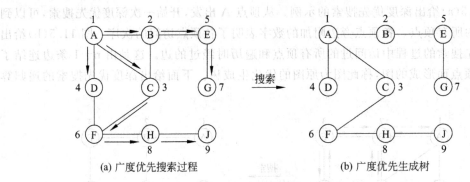

(a) 广度优先搜索过程　　　　　　　　　　(b) 广度优先生成树

图 11.6　广度优先搜索示例

广度优先搜索是一种分层的搜索过程,如图 11.6(a)所示,每向前走一步可能访问一批顶点,不像深度优先搜索那样有往回退的情况。因此,广度优先搜索不是一个递归的过程,其算法也不是递归的。为了实现逐层访问,算法中使用了一个队列,以记忆正在访问的这一层和上一层的顶点,以便于向下一层访问。另外,与深度优先搜索过程一样,为避免重复访问,需要一个辅助数组 visit,给被访问过的顶点加标记。下面给出广度优先搜索的算法。

```
void BFS(int v)
{
    vector<int>visit(vec_vertices.size());    //辅助数组,标识已访问过的顶点
    visit[v]=1;
    cout<<v<<"\t"<<vec_vertices[v].GetNodeDiscript()<<endl;
    queue<int>q;                              //STL queue 队列容器
    q.push(v);
    while(!q.empty())
    {
        v=q.front();                          //读队头元素
        q.pop();                              //删除队头元素
        int w=GetFirstNeighbour(v);
        while(w!=-1)
        {
            if(!visit[w])
            {
                cout<<w<<"\t"<<vec_vertices[w].GetNodeDiscript()<<endl;
                visit[w]=1;
                q.push(w);
```

```
            w=GetNextNeighbour(v,w);
        }
    }                                           //while(!q.empty())
}
```

(5) 最小生成树克鲁斯卡尔算法。

基本思想是：设有一个有 n 个顶点的连通网络 N={V,E}，最初先构造一个只有 n 个顶点，没有边的非连通图 T={V,φ}，图中每个顶点自成一个连通分量。当在 E 中选到一条具有最小权值的边时，若该边的两个顶点落在不同的连通分量上，则将此边加入到 T 中；否则将此边舍去，重新选择一条权值最小的边，如此重复下去，直到所有顶点在同一个连通分量上为止。算法简写框架如下所示。

```
1. T<-φ,T 是最小生成树边的集合,初值为空
2. while (T 包含的边少于 n-1,n 是结点数)
3.    从 E 中选一条具有最小权值的边(v,w);
4.    如果(v,w)加到 T 中后不会产生回路,则将(v,w)加入 T,否则放弃(v,w);
5.    从 E 中删去(v,w);
6. end while
```

源代码如下所示。

```
void MyGraph::Kruskal(MyMinSpanTree<DistType>& t)
{
    vector<MyEdge<DistType>> vec;
    int nSize=vec_vertices.size();              //获得顶点数
    for(int i=0; i<nSize; i++)
                    //读邻接矩阵,获得边的向量集合 vector<MyEdge<DistType>>vec
    {
        for(int j=i+1; j<nSize; j++)
        {
            if(vec_matrix[i][j]!=0)
            {
                MyEdge<DistType>e(i,j,vec_matrix[i][j]);
                vec.push_back(e);
            }
        }
    }
    sort(vec.begin(),vec.end());                //边排序,权值由小到大
    typename vector<MyEdge<DistType>>::iterator it=vec.begin();
    int n=0;
    while(n<nSize-1)                            //最小生成树边数=结点数-1,若不满足,则循环
    {
        if(t.Insert(*it))                       //权值从小到大取边
```

```
            {
                n++;
            }
            it++;
        }
    }
```

该函数涉及了 MyGraph 头文件中定义的所有内容,功能是要把符合条件的边加到最小生成树类 MyMinSpanTree 中定义的边向量 vector<MyEdge<DistType>> vec_Edge 中。在 Kruskal 函数中,边集合 vec 经 sort 排序后,保证了 vec 中各边对象按权值由小到大排列,这样可在循环中保证每次取出的都是当前有效的最小边,因此只需判定该边是否与已符合条件的最小生成树边集合构成回路即可,若构成回路,则必须把该边舍去。示例中当前边是 *it,有一行语句 t.Insert(*it),因此一定是在 MyMinSpanTree 的 Insert 函数中判定 *it 的回路问题。相关代码如下。

```
bool MyMinSpanTree::IsCircuit(int from,int to)
    //从 from 结点开始,利用广度优先搜索判定 to 是否是其结点。若是,则构成回路,舍去 from~to 边
{
    bool bExist=false;
    vector<int>vMark(vertices);
    typename vector<MyEdge< DistType >>::iterator it=find_if(vec_Edge.begin(),vec_Edge.end(),NodeCompare<int>(from));
    if(it==vec_Edge.end())
        return false;

    vMark[from]=1;
    queue<int>q;
    q.push(from);

    while(!q.empty())
    {
        int v=q.front();
        if(v==to)
        {
            bExist=true;
            break;
        }
        q.pop();

        it=find_if(vec_Edge.begin(),vec_Edge.end(),NodeCompare<int>(v));
        while(it!=vec_Edge.end())
        {
            int start=(*it).GetHead();
```

```cpp
            int end=(*it).GetTail();
            if(vMark[start]==0)
            {
                q.push(start);
                vMark[start]=1;
            }
            if(vMark[end]==0)
            {
                q.push(end);
                vMark[end]=1;
            }
            it++;
            it=find_if(it++,vec_Edge.end(),NodeCompare<int>(v));
        }
    }
    return bExist;
}

bool MyMinSpanTree::Insert(MyEdge<DistType>& e)
{
    int from=e.GetHead();
    int to=e.GetTail();
    bool bState=false;

    if(vec_Edge.empty())                        //加入第1条权值最小的边
    {
        vec_Edge.push_back(e);
        bState=true;
    }
    else
    {
        if(!IsCircuit(from,to))
        //对当前最小边判断是否与已有最小生成树边集合构成回路,若没有,则加入
        {
            vec_Edge.push_back(e);
            bState=true;
        }
    }
    return bState;
}
```

判定回路是在 Insert 函数中调用 IsCircuit 函数实现的。在 IsCircuit 函数中,从 from 结点开始,利用广度优先搜索遍历各结点。若 to 结点在其中,则表明 from～to 边与

已有最小生成树边集合构成回路,所以舍去,否则把 from~to 边加入到最小生成树边集合中。

（6）测试程序。

仍以图 11.5 原型为基础,各边加上权重后如图 11.7 所示。

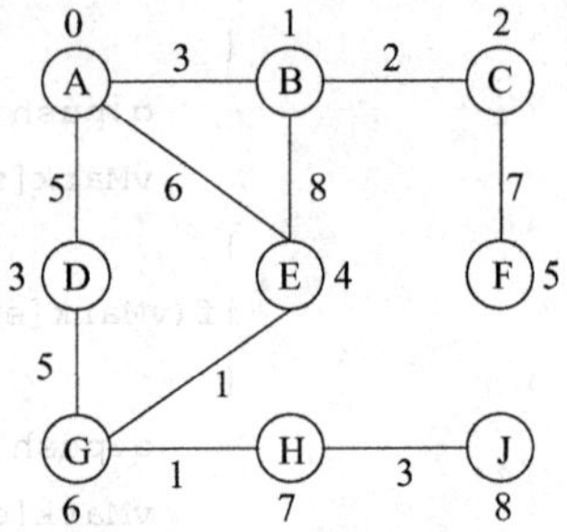

该图结点号依次为 0,1,…,8,对应结点名为 A,B,…,J,边上权重如图 11.7 所示。测试主程序如下所示。

图 11.7　测试用例图

```
int main()
{
    string strNodes="";
    strNodes=strNodes+"0 A 1 B 2 C 3 D 4 E 5 F 6 G 7 H 8 J";
    string strEdge="";
    strEdge=strEdge+"0 3 0 5 6 0 0 0 0 "+ //写成"0 3 0 5 6 0 0 0 0"错误,最后一定要有空格
                   "3 0 2 0 8 0 0 0 0 "+
                   "0 2 0 0 0 7 0 0 0 "+
                   "5 0 0 0 0 0 5 0 0 "+
                   "6 8 0 0 0 0 0 1 0 0 "+
                   "0 0 7 0 0 0 0 0 0 "+
                   "0 0 0 5 1 0 0 1 0 "+
                   "0 0 0 0 0 0 1 0 3 "+
                   "0 0 0 0 0 0 0 3 0";
    MyGraph<MyNode,int>g(strNodes,strEdge,9);
    cout<<"图邻接矩阵是:"<<endl;
    g.Show();

    cout<<"从 0 结点开始 DFS:"<<endl; cout<<"结点号\t 描述"<<endl;
    g.DFS(0);
    cout<<"从 0 结点开始 BFS:"<<endl; cout<<"结点号\t 描述"<<endl;
    g.BFS(0);
    cout<<endl<<endl;

    MyMinSpanTree<int>t(9);            //表明是 9 个结点的最小生成树
    g.Kruskal(t);
    cout<<"最小生成树边集合:"<<endl; cout<<"始结点\t 止结点\t 权值"<<endl;
    t.Show();
    return 0;
}
```

11.3　在 Visual C++ 中应用

Visual C++ 提供了强大的功能,深受编程者喜爱。但其庞大而复杂的类库也使许多人望而却步,那么如果在 Visual C++ 中运用 STL,就能少学许多 MFC 类库,具体体现在如下

方面。

(1) STL string 类可代替 MFC CString。对每种程序开发包而言,字符串编程都是最重要的功能,不同字符串类库尽管功能是非常相似的,但是熟悉函数名写法及参数类型是一项长期的工作。如果熟悉了 STL string 类,那么就省去了这份工作。而且 STL 中还有关于字符串流的操作,功能更强大。

(2) STL 文件操作 ifstream、ofstream、fstream 可以代替关于文件的 CFile 类。

(3) STL 基本序列容器(vector,list,queue,stack)及关联容器(set,map)可代替 MFC 中有关容器的模板类。MFC 中容器模板类有 CArray、CList、CMap、CTypedPtrArray、CTypedPtrList 和 CTypedPtrMap 等。

(4) STL 还能代替许多其他的 MFC 非模板类,如 CByteArray、CDWordArray 和 CPtrList 等,总之那些与 GUI 无关的类都可能由 STL 代替。

(5) 如果在 MFC 中应用了 STL,那么就可以应用强大的 STL 算法。

11.3.1 Scribble 绘图程序

【例 11.14】 MFC Scribble 程序。

Scribble 程序是 MSDN 中一个著名的 MFC 例子,是一个小绘图应用程序,它允许用鼠标绘制徒手画并将图像保存在文件中,可复现。本示例用单文档 MFC+STL 完成了类似的功能。先看一下程序结果。

分析:关键问题有两点:(1)图 11.8 中用鼠标写了"三"字,应该用 3 笔写完,那么在程序中如何标识一笔呢? 我们规定:当鼠标左键按下时是一笔的开始,鼠标左键抬起时是一笔的结束,中间拖动过程完成绘制过程,因此应该响应视图类的 WM_LBUTTONDOWN、WM_LBUTTONUP、WM_MOUSEMOVE;(2)完成一笔绘制应该是多个点的集合,应该有一个基础类能体现出语义"一笔有多个点"。假设类名

图 11.8 Scribble 示例

是 CStroke,Stroke 汉语是"一笔、一划"的意思,内部应该有一个 vector<CPoint>成员变量,CPoint 是 VC 提供的一个系统类,成员变量有 x,y,因此可表示一个点的坐标。

编程步骤如下:

(1) 产生一个单文档工程,工程名为 MyScribble。产生类为 CMyScribbleApp、CMainFrame、CMyScribbleDoc 和 CMyScribbleView。

(2) 新建表示"一笔"的基础类 CStroke。

```
//stroke.h
class CStroke
{
public:
    CStroke();
```

```cpp
        virtual ~CStroke();
public:
        vector<CPoint>m_pointArray;            //一笔有多个点
public:
        void NewPoint(CPoint pt);              //新加一个点
        void DrawStroke(CDC * pDC);            //划这"一笔"
        void LoadStroke(CArchive& ar);         //从文件中读"一笔"
        void SaveStroke(CArchive& ar);         //向文件中存"一笔"
};

//stroke.cpp
CStroke::CStroke(){}
CStroke::~CStroke(){}
void CStroke::NewPoint(CPoint pt)
{
    m_pointArray.push_back(pt);
}
void CStroke::DrawStroke(CDC * pDC)
{
    if(m_pointArray.empty())
        return;
    pDC->MoveTo(m_pointArray[0]);
    for(int i=1; i<m_pointArray.size(); i++)
    {
        pDC->LineTo(m_pointArray[i]);
    }
}
void CStroke::LoadStroke(CArchive& ar)
{
    int nPointSize;
    ar>>nPointSize;
    for(int i=0; i<nPointSize; i++)
    {
        CPoint p;
        ar>>p.x>>p.y;
        m_pointArray.push_back(p);
    }
}
void CStroke::SaveStroke(CArchive& ar)
{
    ar<<m_pointArray.size();
    for(int i=0; i<m_pointArray.size(); i++)
```

```
        {
            CPoint& p=m_pointArray.at(i);
            ar<<p.x<<p.y;
        }
}
```

从 stroke.h 注释中可看出该类基本功能：可以往当前"笔"中新加一点，可以把完整"一笔"存入文件，从文件读出，可以画完整的"一笔"。所有函数都是围绕 vector<CPoint>展开的。以向文件中存一笔为例：先存入该笔画有多少个点，即 vector<CPoint>向量的元素大小，接着遍历该向量，依次存入这些点的坐标。

(3) 在文档类中可读写某"笔画"文件，一般是先按某格式保存文件，然后再读出。由于文件中是多个笔画的集合，因此在文档类 CMyScribbleDoc 中增加 vector<CStroke * > m_strokeList,语义是：一个文件由多"笔"(CStroke)组成，一"笔"(CStroke)由多个点(CPoint)组成，文档中所有函数都是围绕 m_strokeList 展开的。那么向量元素为什么定义成 CStroke *，而不是 CStroke 呢？请看表 11.5 对比说明。

表 11.5 指针与非指针向量差别

代码 1：	代码 2：

Vector<CStroke>v
CStroke c
…   //假设 c 包含 10 000 个点
v.push_back(c);
	vector<CStroke * >v
	CStroke * p=new CStroke()
	… //假设 c 包含 10 000 个点
	v.push_back(p);

左右两侧 c 和 * p 占用的内存空间是一样的，但当压入向量时，左侧是压入一个 c 的副本，占用非常大的内存；而右侧仅压入一个指针，大小 4 个字节，通过指针向量可以方便访问到真实的元素，因此右侧代码效率更高。头文件（黑体部分是新增部分）及部分源文件代码如下所示。

```
//MyScribbleDoc.h
#include"Stroke.h"
class CMyScribbleDoc:public CDocument
{
protected:                      //create from serialization only
    CMyScribbleDoc();
    DECLARE_DYNCREATE(CMyScribbleDoc)

//Attributes
public:
    vector<CStroke * >m_strokeList;        //一文档由多"笔"组成
//Operations
public:
    void NewStroke();                      //往当前"笔"中新增一个结点
```

```cpp
    CStroke * GetNewStroke();                  //获得当前"笔"对象
    void DrawStrokes(CDC * pDC);               //画文件中所有"笔"
    void LoadStrokes(CArchive& ar);            //读文件中所有"笔"
    void SaveStrokes(CArchive& ar);            //存文件中所有"笔"
//Overrides
    //ClassWizard generated virtual function overrides
    //{{AFX_VIRTUAL(CMyScribbleDoc)
    public:
    virtual BOOL OnNewDocument();
    virtual void Serialize(CArchive & ar);
    virtual void DeleteContents();             //删除容器指针元素
    //}}AFX_VIRTUAL
//Implementation
public:
    virtual ~CMyScribbleDoc();
#ifdef _DEBUG
    virtual void AssertValid()const;
    virtual void Dump(CDumpContext & dc)const;
#endif
protected:
//Generated message map functions
protected:
    //{{AFX_MSG(CMyScribbleDoc)
        //NOTE- the ClassWizard will add and remove member functions here.
        //DO NOT EDIT what you see in these blocks of generated code!
    //}}AFX_MSG
    DECLARE_MESSAGE_MAP()
};

//MyScribbleDoc.cpp
void CMyScribbleDoc::NewStroke()               //往当前"笔"中新增一个结点
{
    CStroke * pStrokeItem=new CStroke();
    m_strokeList.push_back(pStrokeItem);
    SetModifiedFlag();                 //Mark the document as having been modified, for
                                       //purposes of confirming File Close.
}
CStroke * CMyScribbleDoc::GetNewStroke()       //获得当前"笔"对象
{
    int nSize=m_strokeList.size();
    return m_strokeList.at(nSize-1);
}
void CMyScribbleDoc::DrawStrokes(CDC * pDC)    //画文件中所有"笔"
```

```cpp
    {
        for(int i=0; i<m_strokeList.size(); i++)
        {
            (m_strokeList.at(i))->DrawStroke(pDC);
        }
    }
    void CMyScribbleDoc::LoadStrokes(CArchive & ar)        //读文件中所有"笔"
    {
        WORD nSize;
        ar>>nSize;
        for(int i=0; i<nSize; i++)
        {
            CStroke * p=new CStroke();
            m_strokeList.push_back(p);
            p->LoadStroke(ar);
        }
    }
    void CMyScribbleDoc::SaveStrokes(CArchive & ar)        //读文件中所有"笔"
    {
        ar<<(WORD)m_strokeList.size();
        for(int i=0; i<m_strokeList.size(); i++)
        {
            CStroke * p=m_strokeList.at(i);
            p->SaveStroke(ar);
        }
    }
    void CMyScribbleDoc::Serialize(CArchive & ar)
    {
        if(ar.IsStoring())
        {
            //TODO: add storing code here
            SaveStrokes(ar);
        }
        else
        {
            //TODO: add loading code here
            LoadStrokes(ar);
        }
    }
    void CMyScribbleDoc::DeleteContents()
    {
        //TODO: Add your specialized code here and/or call the base class
```

```cpp
    for(int i=m_strokeList.size()-1; i>=0; i--)
    {
        delete m_strokeList.at(i);
    }
    CDocument::DeleteContents();
}
```

函数 NewStroke 及 GetNewStroke 这两个函数是与视图类 WM_LBUTTONUP、WM_MOUSEMOVE 消息响应相关联的,希望同学们加以对比分析。LoadStrokes 函数是在打开文件,读文件时起作用的。DrawStrokes 函数是当读完"笔画"文件后第一次显示在视图中,或者当视窗大小发生变化时,或者移去被覆盖窗口时起作用的,当然它是与视图类 OnDraw 函数关联的。

(4) 在视图类 CMyScribbleView 中,增加 WM_LBUTTONDOWN、WM_LBUTTONUP、WM_MOUSEMOVE 响应函数,同时增加成员变量 CPoint m_prev,这是由于在响应 WM_LBUTTONUP、WM_MOUSEMOVE 消息时,需要连续划线,必须保持住前一点。下面列出了头文件及源程序主要的函数源码。头文件中黑体部分是新增部分。

```cpp
//MyScribbleView.h
class CMyScribbleView : public CView
{
protected:                                     //create from serialization only
    CMyScribbleView();
    DECLARE_DYNCREATE(CMyScribbleView)

//Attributes
public:
    CMyScribbleDoc * GetDocument();
    CPoint m_prev;
//Operations
public:
    void DrawLineTo(CPoint pt);
//Overrides
    //ClassWizard generated virtual function overrides
    //{{AFX_VIRTUAL(CMyScribbleView)
    public:
    virtual void OnDraw(CDC * pDC);            //overridden to draw this view
    virtual BOOL PreCreateWindow(CREATESTRUCT& cs);
    protected:
    virtual BOOL OnPreparePrinting(CPrintInfo * pInfo);
    virtual void OnBeginPrinting(CDC * pDC,CPrintInfo * pInfo);
    virtual void OnEndPrinting(CDC * pDC,CPrintInfo * pInfo);
    //}}AFX_VIRTUAL
```

```cpp
//Implementation
public:
    virtual ~CMyScribbleView();
#ifdef _DEBUG
    virtual void AssertValid()const;
    virtual void Dump(CDumpContext& dc)const;
#endif
protected:
//Generated message map functions
protected:
    //{{AFX_MSG(CMyScribbleView)
    afx_msg void OnLButtonDown(UINT nFlags,CPoint point);
    afx_msg void OnLButtonUp(UINT nFlags,CPoint point);
    afx_msg void OnMouseMove(UINT nFlags,CPoint point);
    //}}AFX_MSG
    DECLARE_MESSAGE_MAP()
};

//MyScribbleView.cpp 中主要函数,其余函数略去
void CMyScribbleView::OnLButtonDown(UINT nFlags,CPoint point)
{
    //TODO: Add your message handler code here and/or call default
    GetDocument()->NewStroke();                //增加一个新"笔"
    CStroke * pNewStroke=GetDocument()->GetNewStroke();
    pNewStroke->NewPoint(point);               //新"笔"中增加一个点
    m_prev=point;        //设为新笔起点,在 OnLButtonUp、OnMouseMove 连续画线中用到
    SetCapture();
    CView::OnLButtonDown(nFlags,point);
}
void CMyScribbleView::OnLButtonUp(UINT nFlags,CPoint point)
{
    //TODO: Add your message handler code here and/or call default
    if(GetCapture()!=this)
        return;
    CStroke * pNewStroke=GetDocument()->GetNewStroke();
    pNewStroke->NewPoint(point);
    DrawLineTo(point);                         //画新"笔"最后一点
    ReleaseCapture();                          //Release the mouse capture established at
    CView::OnLButtonUp(nFlags,point);
}
void CMyScribbleView::OnMouseMove(UINT nFlags,CPoint point)
{
    //TODO: Add your message handler code here and/or call default
```

```
    if(GetCapture()!=this)
        return;
    CStroke * pNewStroke=GetDocument()->GetNewStroke();
    pNewStroke->NewPoint(point);
    DrawLineTo(point);                              //连续画新"笔"
    m_prev=point;
    CView::OnMouseMove(nFlags,point);
}
void CMyScribbleView::DrawLineTo(CPoint pt)
{
    CClientDC dc(this);
    dc.MoveTo(m_prev);
    dc.LineTo(pt);
}
void CMyScribbleView::OnDraw(CDC * pDC)
{
    CMyScribbleDoc * pDoc=GetDocument();
    ASSERT_VALID(pDoc);
    //TODO: add draw code for native data here
    pDoc->DrawStrokes(pDC);
}
```

从上述事例中可以看出 STL vector 的作用，或许能得出学好 VC 的技巧，那就是学好 VC 关键框架思路，如文档－视结构，消息响应机制，GUI 类，底层的很多基础功能可用 STL 来实现。

11.3.2　数据库操作程序

【例 11.15】　数据库表解析程序。

先看一下程序运行结果，如图 11.9 所示。

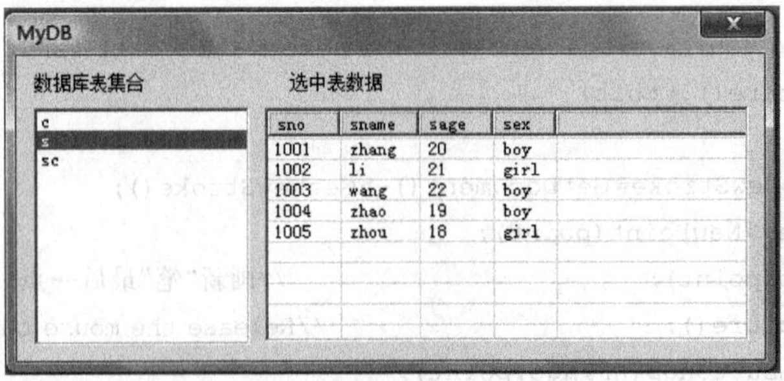

图 11.9　数据库表解析功能示例

完成功能是：初始化时左侧列表框显示某数据库的所有表名。当用鼠标双击该列表框中的某表名时，右侧列表控件在表头处显示该表各字段的属性名称，在数据区显示该表的所有记录信息。

为了简化起见,建立一个 Access 数据库,在库中建立几个表,并在控制面板 ODBC 数据源管理器进行注册,假设名称为 mysrc。由于 MFC 框架产生的应用程序代码都较大,仅把关键代码列出来加以说明。

编程步骤如下：

(1) 产生一个基于对话框的工程,工程名为 MyDB,OLE 自动化选项要选上,生成许多相关类,与本事例相关的是 CMyDBApp、CMyDBDlg。注意：CMyDBApp 源文件中一定要包含下述黑体字代码。若没有,是由于产生工程时没有选择 OLE 自动化选项造成的,可直接在源码上添加,并在 stdafx.h 中加入头文件 #include <afxdisp.h>。

```
//MyDB.cpp
BEGIN_MESSAGE_MAP(CMyDBApp,CWinApp)
    ON_COMMAND(ID_HELP,CWinApp::OnHelp)
END_MESSAGE_MAP()
/////////////////////////////////////////////////////////////////////
//CMyDBApp construction
CMyDBApp::CMyDBApp()
{}
/////////////////////////////////////////////////////////////////////
//The one and only CMyDBApp object
CMyDBApp theApp;
/////////////////////////////////////////////////////////////////////
//CMyDBApp initialization
BOOL CMyDBApp::InitInstance()
{
    //Initialize OLE libraries
    if(!AfxOleInit())
    {
        AfxMessageBox(IDP_OLE_INIT_FAILED);
        return FALSE;
    }
    CMyDBDlg dlg;
    m_pMainWnd=&dlg;
    int nResponse=dlg.DoModal();
    return FALSE;
}
```

黑体部分代码必须有,因为本示例中数据库操作采用了 ADO 组件技术,应用程序与该组件通信是通过 OLE 自动化技术实现的,所以必须加载 OLE 自动化库。

(2) 编制数据库操作类 CAdoFunc。

人们常常需要一种简单统一的 API 以使自己的应用能访问、更改各类不同的数据源。一个数据源可能像一个文本文件一样简单,也可能像一个由各种数据库组成的集群那样复杂，甚至可能是将来才被发明的。而且这个 API 也不能预先设定访问和操作数据源的方法。尽管有这样的一些特殊要求，但一个典型的数据源总是一个关系数据库，它符合开放型数据库连接标准，并且使用 SQL 语句进行操作。对此，由微软提供的一个通用解决方案是 OLE DB，它是一组 COM 接口的集合，提供了统一的方法以访问存储在不同信息源中的数据。但是，由于 OLE DB API 是为了给尽可能多的不同应用提供最佳功能而设计的，因此不符合使用简便这一要求。所以需要一个介于 OLE DB 和应用间的桥梁，而 ADO 正是这座桥梁。

ADO 提供了一系列的通用方法以执行下列的顺序操作来访问数据源：

① 连接到一个数据源。可以选择完成或取消对数据源做出的所有改变。
② 执行指定命令以访问数据源，并可以使用可变参数，或做出性能方面的优化。
③ 执行命令。
④ 执行命令后从一张表中的若干记录行得到的数据将会返回至缓冲区中存放，使用户可以轻松地对数据进行检查、更改之类的操作，并在需要的时候利用缓冲区中的数据更新数据源。
⑤ 提供一个通用的方法检测进行连接或执行命令时的错误。

应用 ADO 首先要引入 ADO 类，本示例是在 stdafx.h 中引入的，加入如下代码。

```
#import "c:\\program files\\common files\\system\\ado\\msado15.dll" no_namespace
```

rename("EOF","adoEOF")，这样就可以用 3 个智能指针了：_ConnectionPtr、_RecordsetPtr 和 _CommandPtr。

```
//AdoFunc.h 头文件
class CAdoFunc
{
public:
    CAdoFunc();
    virtual ~CAdoFunc();
private:
    _ConnectionPtr   m_pConnect;
    _CommandPtr      m_pCommand;
    _RecordsetPtr    m_pRecordSet;
public:
    //Operate function
    BOOL Connect(CString connect,CString user,CString pwd);   //连接数据源
    void Close();                                             //关闭数据源
    BOOL GetTablesName(vector<CString>& vecTable);            //获得表名集合
    BOOL Execute(CString strSQL);                             //执行 4 种常用 SQL 语句
    BOOL IsEmpty();                                           //select 后记录集是否空
    BOOL IsEOF();                                             //是否记录集尾
    void MoveFirst();                                         //移到第一个记录
    void MoveNext();                                          //移到下一个记录
```

```cpp
    void MoveLast();                              //移到最后一个记录
    int GetFieldsNum();                           //获得字段数目
    CString GetFieldName(int nIndex);             //获得某字段名称
    int GetFieldType(int nIndex);                 //获得某字段类型
    int GetFieldLimit(int nIndex);                //获得某字段最大内存长度(字节)
    CString GetFieldValue(int nIndex);            //获得某字段值
};
//AdoFunc.cpp源文件
CAdoFunc::CAdoFunc(){}
CAdoFunc::~CAdoFunc(){}
BOOL CAdoFunc::Connect(CString connect,CString user,CString pwd)    //连接数据源
{
    _bstr_t sconnect(connect);
    _bstr_t suser(user);
    _bstr_t spwd(pwd);
    try
    {
        HRESULT hr=m_pConnect.CreateInstance(__uuidof(Connection));
        if(SUCCEEDED(hr))
        {
            if(SUCCEEDED(m_pConnect->Open(sconnect,suser,spwd,adModeUnknown)))
            {
                hr=m_pCommand.CreateInstance(__uuidof(Command));
                if(SUCCEEDED(hr))
                    m_pCommand->ActiveConnection=m_pConnect;
                return TRUE;
            }
        }
    }
    catch(_com_error e)
    {
        AfxMessageBox("数据库连接错误!");
        return FALSE;
    }
    return TRUE;
}
void CAdoFunc::Close()                            //关闭数据源
{
    try
    {
        m_pCommand.Release();
        m_pCommand=NULL;
        m_pConnect->Close();
```

```cpp
            m_pConnect.Release();
            m_pConnect=NULL;
        }
        catch(_com_error e)
        {
            AfxMessageBox("数据库关闭失败");
        }
    }
    BOOL CAdoFunc::GetTablesName(vector<CString>& vecTable)    //获得表名集合
    {
        if(m_pConnect==NULL)return FALSE;

        USES_CONVERSION ;
        _RecordsetPtr pSet;
        pSet=m_pConnect->OpenSchema(adSchemaTables);
        while(!(pSet->adoEOF))
        {
            //获取表格
            _bstr_t table_name=pSet->Fields->GetItem("TABLE_NAME")->Value;
            //获取表格类型
            _bstr_t table_type=pSet->Fields->GetItem("TABLE_TYPE")->Value;
            //过滤一下,只输出表格名称,其他的省略
            if(strcmp(((LPCSTR)table_type),"TABLE")==0){
                vecTable.push_back(OLE2T(table_name));
            }
            pSet->MoveNext();
        }
        pSet->Close();
        return true;
    }
    BOOL CAdoFunc::Execute(CString strSQL)           //执行 4 种常用 SQL 语句
    {
        if(m_pRecordSet!=NULL)
        {
            m_pRecordSet->Close();
            m_pRecordSet=NULL;
        }
        _bstr_t bstrSQL(strSQL);
        USES_CONVERSION;
        try
        {
            m_pCommand->CommandText=bstrSQL;
            m_pCommand->CommandType=adCmdText;
```

```
            m_pRecordSet=m_pCommand->Execute(NULL,NULL,adCmdText);
    }
    catch(_com_error e)
    {
        AfxMessageBox("SQL 语句执行错误");
        return FALSE;
    }
    return TRUE;
}
BOOL CAdoFunc::IsEmpty()                        //select 后记录集是否空
{
    if(m_pRecordSet->BOF)
    {
        return TRUE;
    }
    return FALSE;
}

BOOL CAdoFunc::IsEOF()                          //是否记录集尾
{
    return m_pRecordSet->adoEOF;
}
void CAdoFunc::MoveFirst()                      //移到第一个记录
{
    if(m_pRecordSet)
    {
        m_pRecordSet->MoveFirst();
    }
}
void CAdoFunc::MoveNext()                       //移到下一个记录
{
    if(m_pRecordSet)
    {
        m_pRecordSet->MoveNext();
    }
}
void CAdoFunc::MoveLast()                       //移到最后一个记录
{
    if(m_pRecordSet)
    {
        m_pRecordSet->MoveLast();
    }
}
```

```cpp
int CAdoFunc::GetFieldsNum()                    //获得字段数目
{
    //only attach to select SQL string
    if(m_pRecordSet==NULL)
    {
        return-1;
    }
    FieldsPtr fpS=m_pRecordSet->GetFields();
    return fpS->Count;
}
CString CAdoFunc::GetFieldName(int nIndex)      //获得某字段名称
{
    USES_CONVERSION;
    FieldsPtr pFields;
    m_pRecordSet->get_Fields(&pFields);
    FieldPtr pf=pFields->GetItem(_variant_t((long)nIndex));
    _bstr_t bt=pf->GetName();
    CString strFieldName=OLE2T(bt);
    strFieldName.TrimLeft();
    strFieldName.TrimRight();
    return strFieldName;
}
int CAdoFunc::GetFieldType(int nIndex)          //获得某字段类型
{
    if(m_pRecordSet==NULL)
    {
        return INVALID_VALUE;
    }
    FieldsPtr fpS=m_pRecordSet->GetFields();
    CComVariant varIndex(nIndex);
    FieldPtr fp=fpS->GetItem(varIndex);
    return fp->GetType();
}
int CAdoFunc::GetFieldLimit(int nIndex)         //获得某字段最大内存长度(字节)
{
    if(m_pRecordSet==NULL)
    {
        return INVALID_VALUE;
    }
    FieldsPtr fpS=m_pRecordSet->GetFields();
    CComVariant varIndex(nIndex);
    FieldPtr fp=fpS->GetItem(varIndex);
    return fp->GetDefinedSize();
```

```cpp
}
CString CAdoFunc::GetFieldValue(int nIndex)          //获得某字段值
{
    CComBSTR bstrReturn="";
    _variant_t vt;
    vt=m_pRecordSet->GetCollect((_variant_t)(long)nIndex);
    if(vt.vt==VT_NULL)
    {
        bstrReturn=_T("");
        return bstrReturn;
    }
    vt.ChangeType(VT_BSTR,&vt);
    bstrReturn=vt.bstrVal;
    USES_CONVERSION;
    CString strTmp=OLE2T(bstrReturn);
    strTmp.TrimLeft();
    strTmp.TrimRight();
    vt.Clear();
    return strTmp;
}
```

该类定义了数据库连接、查询、遍历及获得字段属性等函数。由于应用了 ADO 动态链接库 msado15.dll，只是通过智能指针 _ConnectionPtr、_CommandPtr、_RecordsetPtr 调用各统一接口，真正工作其实是由 msado15.dll 完成的，因此该类显得很简洁，当然还有许多待完善的地方，希望同学们加以完善。特别是有一个函数 GetTablesName(vector＜CString＞& vecTable)，用来获得数据库表名称集合，它的参数是 vector 引用类型，可以随着数据库中表个数的变化而变化，体现出了 vector 的作用。

（3）编制描述表的类 CMyTable。

描述一个表一般来说要知道表名、字段个数及属性，都有哪些具体的记录。除了表名外，其他参数都是动态变化的，而适应这种动态变化正是 STL 容器的特长。因此按上述语义得到如下所示的头文件及源文件。

```cpp
//MyTable.h 头文件

struct CMyField                         //每个字段有如下属性
{
    CString strFieldName;               //名称
    int nFieldType;                     //类型
    int nMaxLength;                     //最大字节长度
};
class CMyTable
{
```

```cpp
    public:
        CMyTable(CAdoFunc * pAdoFunc,CString strTableName);    //pAdoFunc:ADO对象指针
        virtual~CMyTable();                     //strTableName:表名
    public:
        CAdoFunc * m_pAdoFunc;                  //ADO对象指针
        CString m_strTableName;                 //表名
        vector<CMyField>m_vecField;             //表中各字段属性集合
        deque<deque<CString>>m_vecData;         //select查询后封装集合
    public:                                     //数据类型都是字符串
        BOOL ExecuteQuery(CString strSQL);      //查询函数
        BOOL Execute(CString strSQL);           //其他非select函数:insert、update、delete
    };
    //MyTable.cpp源文件
    CMyTable::CMyTable(CAdoFunc * pAdoFunc,CString strTableName)
    {
        m_pAdoFunc=pAdoFunc;                    //ADO对象指针
        m_strTableName=strTableName;            //表名
    }
    CMyTable::~CMyTable(){}
    BOOL CMyTable::ExecuteQuery(CString strSQL)    //select查询函数
    {
        m_pAdoFunc->Execute(strSQL);
        //获得各字段属性向量集
        CMyField myfield;
        int nSize=m_pAdoFunc->GetFieldsNum();
        for(int i=0; i<nSize; i++)
        {
            myfield.strFieldName=m_pAdoFunc->GetFieldName(i);
            myfield.nFieldType=m_pAdoFunc->GetFieldType(i);
            myfield.nMaxLength=m_pAdoFunc->GetFieldLimit(i);
            m_vecField.push_back(myfield);
        }
        if(m_pAdoFunc->IsEmpty())
            return FALSE;
        //获得数据二维向量
        deque<CString>d;
        m_pAdoFunc->MoveFirst();
        while(!m_pAdoFunc->IsEOF())
        {
            d.clear();
            for(int j=0; j<nSize; j++)
            {
```

```
            CString strValue=m_pAdoFunc->GetFieldValue(j);
            d.push_back(strValue);
        }
        m_vecData.push_back(d);
        m_pAdoFunc->MoveNext();
    }
    return TRUE;
}
BOOL CMyTable::Execute(CString strSQL)            //insert、update、delete 功能
{
    m_pAdoFunc->Execute(strSQL);
    return TRUE;
}
```

应用该类前，数据库必须连接上，这样才能通过构造函数 CMyTable（CAdoFunc * pAdoFunc，CString strTableName）建立 CMyTable 有效对象，成员变量数据库对象指针 m_pAdoFunc 才有意义。

成员变量 m_vecField 是 CMyField 的集合。CMyField 是描述字段属性的结构体，可包含字段名称、类型、定义的最大内存字节长度三个属性。由于记录集二维表数据量可能很大，为了避免反复重新分配内存，数据集成员变量没有定义成 vector<vector<CString>>，而是双端队列 deque<deque<CString>>。可以看出目前不论原先数据库中数据定义成何种类型，存入的都是字符串。那么如何恢复成原始数据呢？同学们可以结合字段属性结构体 CMyField 中 nFieldType 的值去解决这一问题。

成员函数 Execute(CString strSQL)主要是用于 insert、update、delete 等 SQL 语句的。ExecuteQuery(CString strSQL)是用于 select SQL 语句的，在该函数内完成了字段属性集及数据集的获得。

（4）完善对话框类 CMyDBDlg。

CAdoFunc、CMyTable 是基础类，CMyDBDlg 是对这两个类的操作，并把操作结果显示在界面上。头文件与源文件如下所示。

```
//MyDBDlg.h 头文件
class CMyDBDlg : public CDialog
{
    DECLARE_DYNAMIC(CMyDBDlg);
//Construction
public:
    CMyDBDlg(CWnd * pParent=NULL);            //standard constructor
    virtual~CMyDBDlg();
//Dialog Data
    //{{AFX_DATA(CMyDBDlg)
    enum{IDD=IDD_MYDB_DIALOG};
```

```cpp
    CListCtrl    m_ctlDataList;                              //数据填充容器
    CListBox     m_ctlTableList;                             //表名填充容器
    //}}AFX_DATA

    //ClassWizard generated virtual function overrides
    //{{AFX_VIRTUAL(CMyDBDlg)
    protected:
    virtual void DoDataExchange(CDataExchange* pDX);    //DDX/DDV support
    //}}AFX_VIRTUAL

//Implementation
public:
    vector<CString>m_vecTable;                               //数据库中表名集合类
    CAdoFunc       m_ado;                                    //ADO 操作类对象
public:
    void FillTableList();                                    //向 listbox 控件添加表名
    BOOL AddItem(int nRow,int nCol,CString strText);
protected:                                                   //向 listctrl 控件添数据函数
    //Generated message map functions
    //{{AFX_MSG(CMyDBDlg)
    virtual BOOL OnInitDialog();                             //完成数据库连接功能
    afx_msg void OnDestroy();                                //完成关闭数据库连接
    afx_msg void OnShowTableData();                          //双击响应函数,用以显示表中数据
    //}}AFX_MSG
    DECLARE_MESSAGE_MAP()
};

//MyDBDlg.cpp 源文件
IMPLEMENT_DYNAMIC(CMyDBDlg,CDialog);

CMyDBDlg::CMyDBDlg(CWnd* pParent /*=NULL*/)
    : CDialog(CMyDBDlg::IDD,pParent)
{}
CMyDBDlg::~CMyDBDlg()
{}
void CMyDBDlg::DoDataExchange(CDataExchange* pDX)
{
    CDialog::DoDataExchange(pDX);
    //{{AFX_DATA_MAP(CMyDBDlg)
    DDX_Control(pDX,IDC_DATA_LIST,m_ctlDataList);
    DDX_Control(pDX,IDC_TABLE_LIST,m_ctlTableList);
    //}}AFX_DATA_MAP
}
```

```cpp
BEGIN_MESSAGE_MAP(CMyDBDlg,CDialog)
    //{{AFX_MSG_MAP(CMyDBDlg)
    ON_WM_DESTROY()
    ON_LBN_DBLCLK(IDC_TABLE_LIST,OnShowTableData)
    //}}AFX_MSG_MAP
END_MESSAGE_MAP()

////////////////////////////////////////////////////////////////////
//self-defined function
void CMyDBDlg::FillTableList()
{
    if(m_vecTable.empty())return;
    m_ctlTableList.ResetContent();
    for(int i=0; i<m_vecTable.size(); i++)
    {
        m_ctlTableList.AddString(m_vecTable[i]);
    }
}
BOOL CMyDBDlg::AddItem(int nRow,int nCol,CString strText)
{
    USES_CONVERSION;
    LVITEM lvItem;
    lvItem.mask     =LVIF_TEXT;
    lvItem.iItem    =nRow;
    lvItem.iSubItem =nCol;
    lvItem.pszText  =(LPTSTR)(LPCTSTR)strText;

    if(nCol==0)
        return m_ctlDataList.InsertItem(&lvItem);
    else
        return m_ctlDataList.SetItem(&lvItem);
    return TRUE;
}

////////////////////////////////////////////////////////////////////
//CMyDBDlg message handlers
BOOL CMyDBDlg::OnInitDialog()
{
    CDialog::OnInitDialog();
    m_ado.Connect("DSN=mysrc","","");           //连接数据源,把某数据库注册成 mysrc
    m_ado.GetTablesName(m_vecTable);            //获得表名向量集合
    FillTableList();                            //填充各表名函数
    m_ctlDataList.SetExtendedStyle(LVS_EX_FULLROWSELECT|LVS_EX_GRIDLINES);
```

```cpp
        return TRUE;                    //return TRUE unless you set the focus to a control
}
void CMyDBDlg::OnDestroy()
{
    CDialog::OnDestroy();

    //TODO: Add your message handler code here
    m_ado.Close();                                          //关闭数据源连接
}
void CMyDBDlg::OnShowTableData()                            //双击响应函数,用以显示表中数据
{
    //TODO: Add your control notification handler code here
    int nSel=m_ctlTableList.GetCurSel();
    if(nSel==LB_ERR)
        return;
    //获取选中表 SQL 语句
    CString strSQL="select * from "+m_vecTable[nSel];
    CMyTable myTable(&m_ado,m_vecTable[nSel]);
    myTable.ExecuteQuery(strSQL);                           //执行查询
    //删除表头及数据
    CHeaderCtrl * phead=m_ctlDataList.GetHeaderCtrl();
    int nCols=phead->GetItemCount();
    if(nCols>=1)
        while(m_ctlDataList.DeleteColumn(0));
    m_ctlDataList.DeleteAllItems();

    //填充表头
    LVCOLUMN  lc;
    lc.mask=LVCF_TEXT|LVCF_WIDTH;
    lc.cchTextMax=100;
    vector<CMyField>& vecField=myTable.m_vecField;
    for(int i=0; i<vecField.size(); i++)
    {
        lc.cx=m_ctlDataList.GetStringWidth(vecField[i].strFieldName)+15;
        lc.pszText=(LPTSTR)(LPCTSTR)vecField[i].strFieldName;

        m_ctlDataList.InsertColumn(i,&lc);
    }

    //填充表中数据
    deque<deque<CString>>& d=myTable.m_vecData;
    for(i=0; i<d.size(); i++)
```

```
            for(int j=0; j<d[0].size(); j++)
            {
                AddItem(i,j,d[i][j]);
            }
```

定义了两个主要的成员变量：CAdoFunc 操作类对象 m_ado 及数据库表向量 m_vecTable。

在 OnInitDialog 函数中，利用 m_ado 连接了数据源，读取并填充了表集合 m_vecTable，根据 m_vecTable 内容调用 FillTableList 成员函数填充了图 11.9 界面左侧的 listbox 控件。

OnShowTableData 是鼠标双击界面左侧 listbox 控件，选中某具体表后，完成右侧 listctrl 控件中显示具体数据的功能。先定义一个局部 CMyTable myTable（&m_ado，m_vecTable[nSel]）对象，m_vecTable[nSel]是选中的表名，含义是 myTable 对象可应用 m_ado 数据库连接，可以对 m_vecTable[nSel]表进行 SQL 语句操作；然后定义 select 查询语句，并通过调用 myTable.ExecuteQuery(strSQL)完成各字段属性及记录集的获得；最后把字段名称显示在 listctrl 控件表头位置上，把记录集数据填充在 listctrl 控件的数据区域中。

11.3.3 文本文件排序、查询

【例 11.16】 文本文件数据排序、查询程序。

CString、CFile 类是 MFC 非常重要的基础类。如果同学们掌握了 STL string、fstream，那么在很大程度上可不必熟悉 CString、CFile 类，但必要的转换还是要掌握的。例如，MFC 许多控件界面填充函数参数都是 CString 类型的，用 string 做参数能填充吗？对话框中若控件值变量是 CString 的，能否用 string 变量接收等。希望同学们带着类似的问题思考本示例。当然，本示例也简单说明了如何在 MFC 应用程序中应用算法的问题。所用文本文件是关于学生基本信息的，包括学号、姓名、年龄和性别 4 项，一行代表一个学生记录，记录中各项间以逗号隔开。先看一下程序运行结果，如图 11.10 所示。

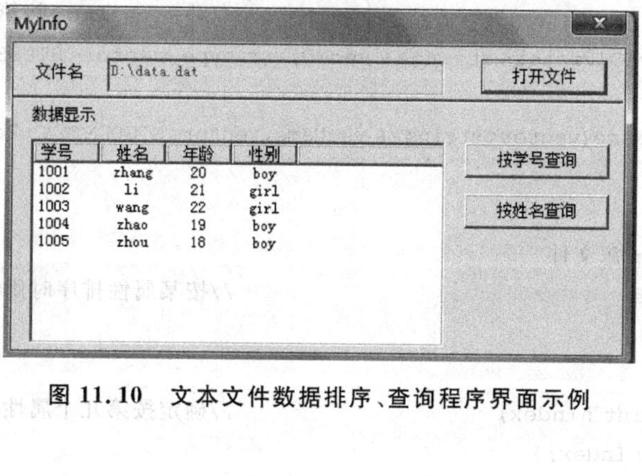

图 11.10 文本文件数据排序、查询程序界面示例

当单击"打开文件"按钮，弹出 CfileDialog 文件选择对话框，选择完文件后，单击"确定"按钮后出现图 11.10 所示的界面。完成的功能如下。

① 排序功能。当按图 11.10 中 listctrl 控件表头中某项时，例如若单击"姓名"时，则表中数据显示按"姓名"升序排列。

② 查询功能。可按"学号"查询，当单击此按钮后，出现学号查询对话框，在编辑框中输入查询的学号。有两种输入方式，对应两种查询类型：一种如"1001,1003,1005"，另一种如"1001-1003"。前者表明查询 1001,1003,1005 学号的学生信息；后者表明是范围查询，查询学号从 1001 到 1003 所有学号的学生信息。

编程步骤如下：

(1) 产生一个基于对话框的工程，工程名为 MyInfo。与本示例相关的类是 CMyInfoApp，CMyInfoDlg。

(2) 产生并编制基础类 CMyProcess，完成具体的排序及查询功能，其头文件及源代码如下所示。

```cpp
//MyProcess.h 头文件
struct STUDENT                                  //学生结构体
{
    string strNO;                               //学号
    string strName;                             //姓名
    int nAge;                                   //年龄
    string strSex;                              //性别
};
class CMyProcess                                //主要处理学生集合的排序,查找功能
{
public:
    CMyProcess(vector<STUDENT>& vecStud);       //以已知学生对象集合建立 CMyProcess 对象
    virtual~CMyProcess();
public:
    vector<STUDENT>& m_vecStud;                 //学生集合引用变量,必须在构造函数完成初始化
    void Sort(int nIndex);                      //对 m_vecStud 按第 nIndex 属性升序排序
    void FindByNO(vector<string>& vecNO,int type,vector<STUDENT>& vResult);
                                                //按学号查询
    void FindByName(vector<string>& vecName,vector<STUDENT>& vResult);   //按姓名查询
};

//MyProcess.cpp 源文件
class MyCompare                                 //按某属性排序时用到函数对象
{
public:
    MyCompare(int nIndex)                       //确定按第几个属性排序
    {m_nIndex=nIndex;}
```

```cpp
public:
    int m_nIndex;
    BOOL operator()(const STUDENT& s1,const STUDENT& s2)const
    {
        BOOL bRet=FALSE;
        switch(m_nIndex)
        {
        case 0:                                    //按学号升序排序
            bRet=(s1.strNO<s2.strNO?TRUE:FALSE);break;
        case 1:                                    //按姓名升序排序
            bRet=(s1.strName<s2.strName?TRUE:FALSE);break;
        case 2:                                    //按年龄升序排序
            bRet=(s1.nAge<s2.nAge?TRUE:FALSE);break;
        case 3:                                    //按性别升序排序
            bRet=(s1.strSex<s2.strSex?TRUE:FALSE);break;
        }
        return bRet;
    }
};
class MyCompare2                                   //按学号查询用到的函数对象
{
public:
    MyCompare2(vector<string>& vecNO,int type):
        m_vecNO(vecNO),m_type(type)
    {}
public:
    int m_type;           //确定查询类型,type=0→{1001,1003,1005};type=1→{1001-1003}
    vector<string>& m_vecNO;                       //待查询学号集合
    BOOL operator()(const STUDENT& s)
    {
        BOOL bRet=FALSE;
        if(m_type==0)              //第1类查询,看s.strNO是否在集合m_vecNO内
            bRet=find(m_vecNO.begin(),m_vecNO.end(),s.strNO)==m_vecNO.end()?TRUE:FALSE;
        if(m_type==1)              //第2类查询,看s.strNO是否在最小学号和最大学号范围内
            bRet=!((s.strNO>=m_vecNO[0] && s.strNO<=m_vecNO[1]));
        return bRet;
    }
};
class MyCompare3                                   //按姓名查询用到的函数对象
{
public:
    MyCompare3(vector<string>& vecName):m_vecName(vecName)
```

```cpp
    {}
public:
    vector<string>& m_vecName;
    BOOL operator()(const STUDENT& s)
    {
        BOOL bRet=FALSE;
        bRet=find(m_vecName.begin(),m_vecName.end(),s.strName)==m_vecName.end()?TRUE:FALSE;
        return bRet;
    }
};
CMyProcess::CMyProcess(vector<STUDENT>& vecStud):m_vecStud(vecStud)
{}
CMyProcess::~CMyProcess()
{}
void CMyProcess::Sort(int nIndex)            //对m_vecStud按第nIndex属性升序排序
{
    sort(m_vecStud.begin(),m_vecStud.end(),MyCompare(nIndex));
}
void CMyProcess::FindByNO(vector<string>& vecNO,int type,vector<STUDENT>& vResult)
                                             //按学号查询
{
    remove_copy_if(m_vecStud.begin(),m_vecStud.end(),back_inserter(vResult),MyCompare2(vecNO,type));
}
void CMyProcess::FindByName(vector<string>& vecName,vector<STUDENT>& vResult)
                                             //按姓名查询
{
    remove_copy_if(m_vecStud.begin(),m_vecStud.end(),back_inserter(vResult),MyCompare3(vecName));
}
```

CMyProcess类完全是一个应用STL的类,成员变量vector<STUDENT>& m_vecStud是引用类型,必须在构造函数中进行初始化,也就是说若应用CMyProcess对象,学生信息集合必须已经存在,CMyProcess类就是关于这些学生信息集的排序和查询。有的同学可能说应该把CMyProcess再细化分成两个类:一个是关于排序,一个是关于查询,这样更符合面向对象思想,当然是可以的。一般来说,相对于MFC类而言,STL类相当于底层。编制应用程序时要尽量确定需要哪些相关的STL基础类,把相同的功能尽量封装在一个类中,不要东一块,西一块。

该类中应用了STL中sort、remove_copy_if、find算法及函数对象MyCompare、MyCompare2、MyCompare3,可以看出在MFC中应用STL并不神秘,方法和编制之前的C++程序是一样的。

(3) 产生按学号查询对话框类 CNOFindDlg，CNameFindDlg。这两个类几乎完全一样，仅以 CNOFindDlg 加以说明。其界面如图 11.11 所示。

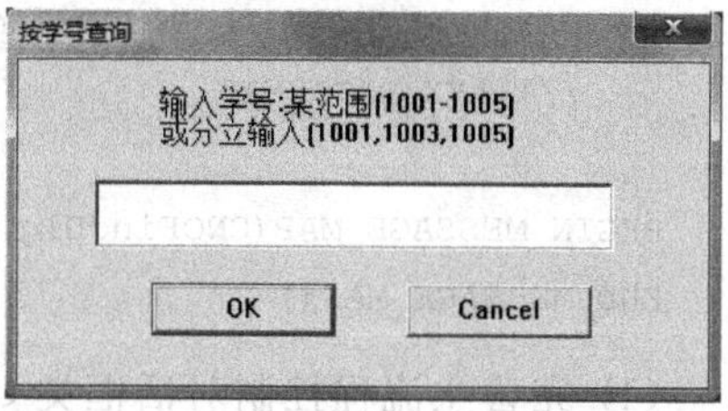

CNOFindDlg 头文件及源文件代码如下所示。

图 11.11　学号查询对话框类

```
//CNOFindDlg.h 头文件
class CNOFindDlg : public CDialog
{
public:
    CNOFindDlg(CWnd * pParent=NULL);       //standard constructor
//Dialog Data
    //{{AFX_DATA(CNOFindDlg)
    enum{IDD=IDD_FIND_DIALOG};
    CString m_strNO;
    //}}AFX_DATA
//Overrides
    //ClassWizard generated virtual function overrides
    //{{AFX_VIRTUAL(CNOFindDlg)
    protected:
    virtual void DoDataExchange(CDataExchange * pDX);  //DDX/DDV support
    //}}AFX_VIRTUAL
//Implementation
protected:
    //Generated message map functions
    //{{AFX_MSG(CNOFindDlg)
        //NOTE: the ClassWizard will add member functions here
    //}}AFX_MSG
    DECLARE_MESSAGE_MAP()
};
//CNOFindDlg.cpp 源文件
CNOFindDlg::CNOFindDlg(CWnd * pParent /* =NULL */)
    : CDialog(CNOFindDlg::IDD,pParent)
{
    //{{AFX_DATA_INIT(CNOFindDlg)
    m_strNO=_T("");
    //}}AFX_DATA_INIT
}
void CNOFindDlg::DoDataExchange(CDataExchange * pDX)
{
    CDialog::DoDataExchange(pDX);
```

```
    //{{AFX_DATA_MAP(CNOFindDlg)
    DDX_Text(pDX,IDC_NO_EDIT,m_strNO);
    //}}AFX_DATA_MAP
}
BEGIN_MESSAGE_MAP(CNOFindDlg,CDialog)
END_MESSAGE_MAP()
```

(4) 完善主流程控制对话框类 CMyInfoDlg。

```
//MyInfoDlg.h 头文件
class CMyInfoDlg : public CDialog
{
//Construction
public:
    CMyInfoDlg(CWnd* pParent=NULL);                //standard constructor
//Dialog Data
    //{{AFX_DATA(CMyInfoDlg)
    enum{IDD=IDD_MYINFO_DIALOG};
    CListCtrl   m_ctlDataList;
    CString     m_strFileName;
    //}}AFX_DATA
    //ClassWizard generated virtual function overrides
    //{{AFX_VIRTUAL(CMyInfoDlg)
    protected:
    virtual void DoDataExchange(CDataExchange* pDX);   //DDX/DDV support
    //}}AFX_VIRTUAL
    vector<STUDENT>m_vecStud;                          //学生向量集合成员变量
    void FillDataList(vector<STUDENT>& vecStud);       //用学生集合数据填充 listctrl 控件
    void GetNOCollect(string strNO,vector<string>& vecNO,int& type);
                                                       //获得待查询学号集合及查询类型
    void GetNameCollect(string strName,vector<string>& vecName); //获得待查询姓名集合
protected:
    //Generated message map functions
    //{{AFX_MSG(CMyInfoDlg)
    virtual BOOL OnInitDialog();
    afx_msg void OnSelectFile();                       //"打开文件"按钮响应函数
    afx_msg void OnFindNo();                           //"按学号查询"按钮响应函数
    afx_msg void OnFindName();                         //"按姓名查询"按钮响应函数
    afx_msg void OnColumnclickDataList(NMHDR* pNMHDR,LRESULT* pResult);
                                                       //控件列排序响应函数
    //}}AFX_MSG
    DECLARE_MESSAGE_MAP()
```

```cpp
};
//MyInfoDlg.cpp 源程序
CMyInfoDlg::CMyInfoDlg(CWnd* pParent /*=NULL*/)
    : CDialog(CMyInfoDlg::IDD,pParent)
{
    //{{AFX_DATA_INIT(CMyInfoDlg)
    m_strFileName=_T("");
    //}}AFX_DATA_INIT
}
void CMyInfoDlg::DoDataExchange(CDataExchange* pDX)
{
    CDialog::DoDataExchange(pDX);
    //{{AFX_DATA_MAP(CMyInfoDlg)
    DDX_Control(pDX,IDC_DATA_LIST,m_ctlDataList);
    DDX_Text(pDX,IDC_FILE_NAME,m_strFileName);
    //}}AFX_DATA_MAP
}
BEGIN_MESSAGE_MAP(CMyInfoDlg,CDialog)
    //{{AFX_MSG_MAP(CMyInfoDlg)
    ON_BN_CLICKED(IDC_FILE_BTN,OnSelectFile)
    ON_BN_CLICKED(IDC_FIND_NO,OnFindNo)
    ON_NOTIFY(LVN_COLUMNCLICK,IDC_DATA_LIST,OnColumnclickDataList)
    ON_BN_CLICKED(IDC_FIND_NAME,OnFindName)
    //}}AFX_MSG_MAP
END_MESSAGE_MAP()
void CMyInfoDlg::FillDataList(vector<STUDENT>& vecStud)
                                                //用学生集合数据填充listctrl控件
{
    m_ctlDataList.DeleteAllItems();
    CString strTmp;;
    for(int nRow=0; nRow<vecStud.size(); nRow++)
    {
        STUDENT& s=vecStud[nRow];
        m_ctlDataList.InsertItem(nRow,s.strNO.c_str());
        m_ctlDataList.SetItem(nRow,1,LVIF_TEXT,s.strName.c_str(),-1,0,0,0);
        strTmp.Format("%d",s.nAge);
        m_ctlDataList.SetItem(nRow,2,LVIF_TEXT,strTmp,-1,0,0,0);
        m_ctlDataList.SetItem(nRow,3,LVIF_TEXT,s.strSex.c_str(),-1,0,0,0);
    }
}                                               //获得待查询学号集合及查询类型
```

```cpp
void CMyInfoDlg::GetNOCollect(string strNO,vector<string>& vecNO,int& type)
{
    int first=0,last=0;
    string s;
    if(strNO.find(",")!=string::npos)              //第1类查询"1001,1003,1005"
    {
        while((last=strNO.find_first_of(",",first))!=string::npos)
        {
            s=strNO.substr(first,last-first);
            vecNO.push_back(s);
            first=last+1;
        }
        s=strNO.substr(first,strNO.length()-first);
        vecNO.push_back(s);
    }
    else
    {
        last=strNO.find("-");
        s=strNO.substr(first,last-first);
        vecNO.push_back(s);
        first=last+1;
        s=strNO.substr(first,strNO.length()-first);
        vecNO.push_back(s);
    }
}
void CMyInfoDlg::GetNameCollect(string strName,vector<string>& vecName)
{                                                  //获得待查询姓名集合
    int first=0,last=0;
    string s;
    while((last=strName.find_first_of(",",first))!=string::npos)
    {
        s=strName.substr(first,last-first);
        vecName.push_back(s);
        first=last+1;
    }
    s=strName.substr(first,strName.length()-first);
    vecName.push_back(s);
}
BOOL CMyInfoDlg::OnInitDialog()
{
```

```cpp
    CDialog::OnInitDialog();
    //填充 listctrl 控件表头内容
    m_ctlDataList.InsertColumn(0,"学号",LVCFMT_CENTER,50);
    m_ctlDataList.InsertColumn(1,"姓名",LVCFMT_CENTER,50);
    m_ctlDataList.InsertColumn(2,"年龄",LVCFMT_CENTER,50);
    m_ctlDataList.InsertColumn(3,"性别",LVCFMT_CENTER,50);
    return TRUE;                   //return TRUE unless you set the focus to a control
}
void CMyInfoDlg::OnSelectFile()                     //"打开文件"按钮响应函数
{
    //TODO: Add your control notification handler code here
    CFileDialog dlg(TRUE);
    if(dlg.DoModal()==IDOK)
    {
        m_strFileName=dlg.GetPathName();            //获得文件名
        UpdateData(FALSE);

        STUDENT stud;
        ifstream in(m_strFileName);                 //读文件形成学生向量集
        while(!in.eof())
        {
            in>>stud.strNO>>stud.strName>>stud.nAge>>stud.strSex;
            m_vecStud.push_back(stud);
        }in.close();
        FillDataList(m_vecStud);                    //用 m_vecStud 填充 listctrl 控件
    }
}
void CMyInfoDlg::OnFindNo()                         //按学号查询
{
    CNOFindDlg dlg;
    if(dlg.DoModal()==IDOK)
    {
        string strNO=dlg.m_strNO;
        int type=0;
        vector<string>vecNO;
        GetNOCollect(strNO,vecNO,type);             //获得待查询学号集合及查询类型
        vector<STUDENT>vResult;                     //初始化查询结果集 vResult 为空
        CMyProcess obj(m_vecStud);
        obj.FindByNO(vecNO,type,vResult);           //完成查询
        FillDataList(vResult);                      //用 vResult 填充 listctrl 控件
    }
}
```

```
}
void CMyInfoDlg::OnFindName()
{
    CNameFindDlg dlg;
    if(dlg.DoModal()==IDOK)
    {
        string strName=dlg.m_strName;
        vector<string>vecName;
        GetNameCollect(strName,vecName);        //获得待查询信号集合
        int nSize=vecName.size();
        vector<STUDENT>vResult;                 //初始化查询结果集 vResult 为空
        CMyProcess obj(m_vecStud);
        obj.FindByName(vecName,vResult);        //完成查询
        FillDataList(vResult);                  //用 vResult 填充 listctrl 控件
    }
}
void CMyInfoDlg::OnColumnclickDataList(NMHDR * pNMHDR,LRESULT * pResult)
                                                //表头列排序响应函数
{
    NM_LISTVIEW * pNMListView=(NM_LISTVIEW * )pNMHDR;
    int pos=pNMListView->iSubItem;              //获得按哪列进行排序
    CMyProcess obj(m_vecStud);
    obj.Sort(pos);                              //完成排序
    FillDataList(m_vecStud);                    //用 m_vecStud填充 listctrl 控件
}
```

在"打开文件"按钮响应函数 OnSelectFile 中,通过 CFileDialog 对象获得了打开的文件名 m_strFileName,它是 CString 类型的。紧接着用 istream in(m_strFileName)建立输入文件流,说明 CString 类型参数可直接应用于 STL 流对象构造函数参数中。在按学号查询对话框类中,成员变量 m_strNO 是字符串类型的。在按对号查询响应函数 OnFindNo 中,先出现按学号查询对话框,当输入学号信息,单击"确定"按钮后,有一行语句 string strNO=dlg.m_strNO,说明 string 类型变量可以直接等于 CString 变量,但是 CString 变量不能直接等于 string 变量。在填充 listctrl 控件函数 FillDataList 函数中有一行语句 m_ctlDataList.InsertItem(nRow,s.strNO.c_str()),即填充第 nRow 列表头内容,该函数要求第二个参数是 CString 类型或 LPSTR 类型,而 s.strNO 是 string 类型,不能直接应用,必须调用 c_str 函数,即 s.strNO.c_str()才是可以的。

11.3.4 基于配置文件的查询程序

【例 11.17】 基于配置文件的通用查询程序。

配置文件采用 ini 格式,例如某查询配置文件 MyConfig.ini 内容如表 11.6 所示。

表 11.6 MyConfig.ini 文件内容

```
[mydatabase]
connect=dsn=mysrc
user=
pwd=

[mycontent]
学生基本信息查询=按学号查询,按学号范围查询,按姓名查询
学生成绩查询=按学号查询,按姓名查询

[学生基本信息查询]
按学号查询=select * from s where sno='[]';输入学号
按学号范围查询=select * from s where sno between '[]' and '[]';起始学号-结束学号
按姓名查询=select * from s where name='[]';输入姓名

[学生成绩查询]
按学号查询=select * from sc where sno='[]';输入学号
按姓名查询=select sname,cno,grade from s,sc where s.sno=sc.sno and sname='[]';输入姓名
```

通过配置文件,可以得出查询的全部相关内容:①是基于数据库的查询,[mydatabase]中封装了数据库连接信息。获得了连接串 connect,用户名 user,密码 pwd 后,就一定能够与数据库相连。②[mycontent]中定义了查询分类,以及每个分类中都有哪些具体的查询描述。示例中定义了两类查询:学生基本信息查询及学生成绩查询。学生基本信息查询中又有按学号查询、按学号范围查询、按姓名查询三个具体查询的描述。学生成绩查询中又有按学号查询,按姓名查询两个具体查询的描述。③其他的 section,如[学生基本信息查询]、[学生成绩查询]等是什么含义呢?它不是随便填充的,是与[mycontent]相关联的,[mycontent]中的每一个 key 都可以作为一个新的 section 出现,[mycontent]中的每一个 key 包含的多个字符串(按逗号拆分)都可以作为新 section 的 key。新 section 中 value 部分包含了两层含义,由分号相隔,分号前面是查询的 SQL 语句,分号后面是一个由"-"相隔的字符串,是对 SQL 语句中"[]"填充信息的描述说明,SQL 语句中有多少"[]",就对应多少个描述说明。例如"按学号范围查询=select * from s where sno between '[]' and '[]',起始学号-结束学号",含义是 SQL 查询语句是 select * from s where sno between '[]' and '[]',第 1 个"[]"需填充的是起始学号,第 2 个"[]"需填充的是结束学号。

再强调一遍,该配置文件中[mydatabase]、[mycontent]这两个 section 是必不可少的,其他的 section 是由[mycontent]推导并加以丰富得出的。应用程序查询内容随配置文件内容变化而变化,那么如何适应这种动态变化?毫无疑问,STL 是适应这种动态变化较有力的工具。编程完全是针对配置文件,配置文件内容描述了要实现的内容,程序界面只不过是配置文件的一种映射,一种反应方式。

若配置文件是 MyConfig.ini,数据库采用 Access 数据库,包含:学生表 s(sno:学号,sname:姓名,sage:年龄,sex:性别),课程表 c(cno:课号,cname:课名),成绩表 sc(sno:学号,cno:课号,grade:成绩)。程序执行界面示例如图 11.12 和图 11.13 所示。

  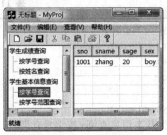

(a) 按学号查询主界面　　　(b) 按学号查询对话框　　　(c) 按学号查询结果

图 11.12　通用查询界面示例 1

(a) 按学号范围查询主界面　　(b) 按学号范围查询对话框　　(c) 按学号范围查询结果

图 11.13　通用查询界面示例 2

当选择学生基本信息查询中按学号查询图 11.12(a)时，弹出查询对话框如图 11.12(b)所示，其中有一个查询输入条件编辑框，输入并单击"确定"按钮后，查询结果如图 11.12(c)所示。当选择学生基本信息查询中按学号范围查询图 11.13(a)时，弹出查询对话框如图 11.13(b)所示，其中有两个查询输入条件编辑框，输入并单击"确定"按钮后，查询结果如图 11.13(c)所示。

通过这两个界面示例，得出基于配置文件的通用查询程序特点如下所示。

(1) 主界面风格一致。左侧是树型视图，显示查询分类及每个分类中具体的查询描述；右侧是列表视图。当然这两个视图显示内容都是动态填充的，是非静态的。

(2) 查询对话框风格一致，查询条件所需输入控件都是动态产生的。本示例中都是编辑控件，若需要其他输入控件，比如时间控件、下拉控件等，只需要丰富配置文件中相应的具体信息就可以了。

编程步骤如下所示。

(1) 产生一个单文档工程，名称为 MyProj，视图类基类选为 ClistView，得到相关类为 CMyProjApp、CMainFrame、CMyProjView 和 CMyProjDoc。

(2) 产生一个新的 MFC 类 CContentView，基类是 CTreeView，用于显示查询分类及每个分类下的具体查询描述。

(3) 产生一个新的普通类 CFindPara，内容如下所示。

```
class CFindPara
{
```

```cpp
public:
    string m_strSection;                              //查询所属 section
    string m_strKey;                                  //查询所属 key
    string m_strSQL;                                  //section-key 对应的 SQL 语句
    vector<string> m_vecDesc;                         //条件描述串
    CFindPara(string section,string key,string strText)
                                                      //strText 中包含 SQL 语句及条件描述串
    {                                                 //必须拆分
        m_strSection=section;
        m_strKey=key;
        int pos=strText.find(";");
        m_strSQL=strText.substr(0,pos);               //第 1 个分号前是 SQL 语句
        pos++;
        strText=strText.substr(pos,strText.length()-pos);
        int first=0,last=0;
        string s;                                     //拆分并获得各个条件描述串
        while((last=strText.find_first_of("-",first))!=string::npos)
        {
            s=strText.substr(first,last-first);
            m_vecDesc.push_back(s);
            first=last+1;
        }
        s=strText.substr(first,strText.length()-first);
        m_vecDesc.push_back(s);
    }
    string MergeSQL(vector<string>& para)             //m_strSQL 与 para 向量合并,得到完整的 SQL 语句
    {
        string s=m_strSQL;
        for(int i=0; i<para.size(); i++)
        {
            int pos=s.find("[]");
            s.replace(pos,2,para[i]);
        }
        return s;
    }
    virtual ~CFindPara(){}
};
```

这是一个查询条件与参数的类,截取表 11.6 中部分内容如表 11.7 所示。

表 11.7 表 11.6 中部分截取内容

[学生基本信息查询]
按学号查询=select * from s where sno='[]';输入学号
按学号范围查询=select * from s where sno between '[]' and '[]';起始学号-结束学号

例如，当执行按学号范围查询时，成员变量 m_strSection＝"学生基本信息查询"，m_strKey＝"按学号范围查询"，m_strSQL＝"select * from s where sno between '[ ]' and '[ ]'"。按表意形式来说，vector＜string＞m_vecDesc＝{"起始学号"，"结束学号"}。

也就是说，该类封装了完成查询所必需的各个参数，是对配置文件相关内容的进一步分类及解析。构造函数 CFindPara(string section, string key, string strText)主要完成了4个成员变量的初始化，举例来说，section 相当于"学生基本信息查询"，key 相当于"按学号范围查询"，strText 相当于"select * from s where sno between '[ ]' and '[ ]'"，起始学号-结束学号"。因此必须对 strText 先按逗号拆分，获得 SQL 语句并给成员变量 m_str SQL 赋值；再对"起始学号-结束学号"按减号拆分，给 m_vecDesc 向量赋值。

由于 m_strSQL 是不完整的 SQL 语句，因此必须把待输入的信息值封装成 vector＜string＞类型的向量，在 MergeSQL(vector＜string＞& para)函数中依次填充 SQL 语句中"[ ]"部分，形成完整的 SQL 语句。

（4）添加读 ini 配置文件类及数据库操作类。

只要把在【例11.3】中有关配置文件的类或结构体，【例11.15】中有关数据库操作的类或结构体添加进来就可以了。从【例11.3】中移植 class MySection、class MySectionCollect、class ReadIni，从【例11.15】中移植 class CAdoFunc、struct CMyField、class CMyTable。一般来说，只要是重复的功能，同学们不要每次都重新编制，尽量移植以前的代码，从中可以懂得原先编制的类有哪些不足，对理解面向对象思想有非常大的促进作用，这也是本示例中要强调的一个主要问题。

（5）编制动态查询对话框类 CFindDlg。

在资源管理器中建立一个图 11.14 所示的对话框。

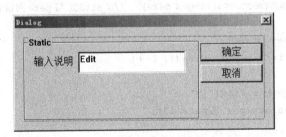

图 11.14　动态查询对话框资源管理器界面

由于查询界面是动态生成的，因此在资源管理器中的界面与实际运行界面是不同的，基本算法是根据查询参数 CFindPara 对象中的 m_vecDesc 字符串向量动态生成界面，若其长度为 n，则添加 n 个编辑框，并在其左侧根据 m_vecDesc 向量内容显示输入说明。这就涉及添加的 CStatic、CEdit 控件的位置，为了方便定位，在资源管理器中加入了两个定位控件 CStatic 及 CEdit。

头文件及源文件代码如下所示。

//FindDlg.h 头文件
class CFindDlg : public CDialog

```cpp
{
//Construction
public:
    CFindDlg(CFindPara * pFindPara,CWnd * pParent=NULL);    //与查询参数对象关联
//Dialog Data
    //{{AFX_DATA(CFindDlg)
    enum{IDD=IDD_FIND_DLG};
    CStatic   m_ctlDesc;
    CEdit     m_ctlFactor;
    CButton   m_ctlSubTitle;
    //}}AFX_DATA
    vector<string> m_vecResult;                             //n个条件的值按向量保存
//Overrides
    //ClassWizard generated virtual function overrides
    //{{AFX_VIRTUAL(CFindDlg)
    protected:
    virtual void DoDataExchange(CDataExchange * pDX);       //DDX/DDV support
    //}}AFX_VIRTUAL
    CString   m_strSection;                                 //查询所属 section,用于界面显示
    CString   m_strKey;                                     //查询所属 key,用于界面显示
    CFindPara * m_pFindPara;                                //与已经定义好的查询参数对象建立关联
    vector<CStatic * > m_vecStatic;                         //动态产生的 CStatic 窗口集合
    vector<CEdit * > m_vecEdit;                             //动态产生的 CEdit 窗口集合
//Implementation
protected:
    //Generated message map functions
    //{{AFX_MSG(CFindDlg)
    virtual BOOL OnInitDialog();                            //根据查询参数对象动态产生 CStatic,CEdit 窗口
    afx_msg void OnDestroy();                               //销毁动态产生的 CStatic,CEdit 窗口
    virtual void OnOK();                                    //收集动态产生的各个 CEdit 中的值
    //}}AFX_MSG
    DECLARE_MESSAGE_MAP()
};

//FindDlg.cpp 源文件
#define MYSTATIC 5000
#define MYEDIT    6000
CFindDlg::CFindDlg(CFindPara * pFindPara,CWnd * pParent /* =NULL * /)
    : CDialog(CFindDlg::IDD,pParent)                        //与查询参数对象关联
{
    m_pFindPara=pFindPara;
}
void CFindDlg::DoDataExchange(CDataExchange * pDX)
```

```cpp
{
    CDialog::DoDataExchange(pDX);
    //{{AFX_DATA_MAP(CFindDlg)
    DDX_Control(pDX,IDC_POS_STATIC,m_ctlDesc);
    DDX_Control(pDX,IDC_FACTOR_EDIT,m_ctlFactor);
    DDX_Control(pDX,IDC_SUB_TITLE,m_ctlSubTitle);
    //}}AFX_DATA_MAP
}

BEGIN_MESSAGE_MAP(CFindDlg,CDialog)
    //{{AFX_MSG_MAP(CFindDlg)
    ON_WM_DESTROY()
    //}}AFX_MSG_MAP
END_MESSAGE_MAP()

BOOL CFindDlg::OnInitDialog()              //根据查询参数对象动态产生CStatic,CEdit窗口
{
    CDialog::OnInitDialog();

    //TODO: Add extra initialization here
    RECT pr1;
    RECT pr2;
    RECT r;
    RECT r2;

    GetWindowRect(&pr1);
    GetClientRect(&pr2);
    m_ctlDesc.GetWindowRect(&r);
    m_ctlFactor.GetWindowRect(&r2);

    int shiftX=pr1.left,shiftY=pr1.bottom-pr2.bottom;
    r.left=r.left-shiftX; r.top=r.top-shiftY;
    r.right=r.right-shiftX; r.bottom=r.bottom-shiftY;
    r2.left=r2.left-shiftX; r2.top=r2.top-shiftY;
    r2.right=r2.right-shiftX; r2.bottom=r2.bottom-shiftY;

    DWORD dwStyle=m_ctlDesc.GetStyle();
    DWORD dwStyle2=m_ctlFactor.GetStyle();
    for(int i=0; i<m_pFindPara->m_vecDesc.size(); i++)
    {
        CStatic * pStatic=new CStatic();
        pStatic->Create(m_pFindPara->m_vecDesc[i].c_str(),dwStyle,r,this,MYSTATIC+i);
        pStatic->ShowWindow(SW_SHOW);
```

第 11 章 STL 应用

```
            m_vecStatic.push_back(pStatic);

            CEdit * pEdit=new CEdit();
            pEdit->Create(dwStyle2|WS_BORDER,r2,this,MYEDIT+i);
            pEdit->ShowWindow(SW_SHOW);
            m_vecEdit.push_back(pEdit);

            r.top+=35; r.bottom+=35;
            r2.top+=35; r2.bottom+=35;
        }
        SetWindowText(m_pFindPara->m_strSection.c_str());   //设置窗口标题,即查询所属 section
        m_ctlSubTitle.SetWindowText(m_pFindPara->m_strKey.c_str());
                                                            //设置子标题,即查询所属 key
        return TRUE;
    }
    void CFindDlg::OnDestroy()                  //销毁动态产生的 CStatic,CEdit 窗口
    {
        CDialog::OnDestroy();
        if(!m_vecStatic.empty())
        {
            for(int i=m_vecStatic.size()-1; i>=0; i--)
            {
                m_vecStatic[i]->DestroyWindow();
                delete m_vecStatic[i];
                m_vecStatic[i]=NULL;
                m_vecEdit[i]->DestroyWindow();
                delete m_vecEdit[i];
                m_vecEdit[i]=NULL;
            }
        }
    }
    void CFindDlg::OnOK() //收集动态产生的各个 CEdit 中的值,保存至 vector<string>m_vecResult
    {
        CString s;
        string sResult;
        for(int i=0; i<m_vecEdit.size();i++)
        {
            m_vecEdit[i]->GetWindowText(s);
            s.TrimLeft();
            s.TrimRight();
            sResult=s;
            m_vecResult.push_back(sResult);
```

```
    }
    CDialog::OnOK();
}
```

构造函数 CFindDlg 中初始化 CFindPara 查询参数对象指针 m_pFindPara，表明在建立 CFindDlg 对象之前，CFindPara 对象必须存在。在 OnInitDialog 函数中，根据 m_pFindPara 中 m_vecDesc 向量动态建立 CStatic 控件，用于输入描述；动态建立 CEdit 控件，用于信息输入。OnOK 是"确定"按钮响应函数，动态收集各 CEdit 控件的输入值，添加到向量 m_vecResult 中。OnDestroy 函数销毁动态产生的 CStatic、CEdit 窗口资源。

(6) 添加劈分窗口。

即左侧是树型视图，右侧是列表视图。打开 mainfrm.h，增加一成员变量 CSplitterWnd m_wndSplitter，之后再增加并修改虚函数 OnCreateClient。

```
BOOL CMainFrame::OnCreateClient(LPCREATESTRUCT lpcs,CCreateContext * pContext)
{
    m_wndSplitter.CreateStatic(this,1,2);
    m_wndSplitter.CreateView(0,0,RUNTIME_CLASS(CContentView),CSize(100,0),pContext);
    m_wndSplitter.CreateView(0,1,pContext->m_pNewViewClass,CSize(0,0),pContext);
    return TRUE;
}
```

(7) 树型视图 CContentView 类、列表视图 CMyProjView 类与文档类关联及相应功能。

对 CMyProjView 而言，主要是建立与文档类关联，是通过重载虚函数 OnInitialUpdate 函数实现的，其他都不用变化，如下所示。

```
void CMyProjView::OnInitialUpdate()    //在此函数内设置文档 CMyProjDoc 中 m_pDataView 值
{
    CListView::OnInitialUpdate();
    CListCtrl & listCtl=GetListCtrl();
    listCtl.SetExtendedStyle(LVS_EX_FULLROWSELECT|LVS_EX_GRIDLINES|
                             LVS_EX_HEADERDRAGDROP);
    CMyProjDoc * pDoc=GetDocument();
    pDoc->m_pDataView=this;
}
```

对 CContentView 而言，要比 CMyProjView 复杂，一是建立与文档类关联，二是要初始化即填充树型视图，三是要响应查询叶结点鼠标双击响应函数，其头文件及源文件内容如下所示。

```
//ContentView.h 头文件
class CContentView : public CTreeView
{
```

```cpp
protected:
    CContentView();    //protected constructor used by dynamic creation
    ~CContentView();
    DECLARE_DYNCREATE(CContentView)
    public:
    virtual void OnInitialUpdate();
                        //在此函数内设置文档 CMyProjDoc 中 m_pContentView 值并填充
    afx_msg int OnCreate(LPCREATESTRUCT lpCreateStruct);
                        //在此函数内修改 TreeView 显示风格
    afx_msg void OnDblclk(NMHDR * pNMHDR,LRESULT * pResult);   //双击树叶结点响应函数
    DECLARE_MESSAGE_MAP()
};

//ContentView.cpp 源文件
IMPLEMENT_DYNCREATE(CContentView,CTreeView)
CContentView::CContentView()
{}
CContentView::~CContentView()
{}


BEGIN_MESSAGE_MAP(CContentView,CTreeView)
    //{{AFX_MSG_MAP(CContentView)
    ON_NOTIFY_REFLECT(NM_DBLCLK,OnDblclk)
    ON_WM_CREATE()
    //}}AFX_MSG_MAP
END_MESSAGE_MAP()

int CContentView::OnCreate(LPCREATESTRUCT lpCreateStruct)
{
    lpCreateStruct->style|=(TVS_HASLINES|TVS_HASBUTTONS|TVS_SHOWSELALWAYS);
    if(CTreeView::OnCreate(lpCreateStruct)==-1)
        return-1;
    return 0;
}
void CContentView::OnInitialUpdate()
                    //在此函数内设置文档 CMyProjDoc 中 m_pContentView 值并填充
{
    CTreeView::OnInitialUpdate();
    CMyProjDoc * pDoc=(CMyProjDoc *)GetDocument();
    pDoc->m_pContentView=this;
    pDoc->FillContentView();
```

```cpp
}
void CContentView::OnDblclk(NMHDR * pNMHDR,LRESULT * pResult)    //双击树叶结点响应函数
{                                                                 //只有叶结点是功能结点
    //TODO: Add your control notification handler code here
    CTreeCtrl& treeCtrl=GetTreeCtrl();
    HTREEITEM hItem=treeCtrl.GetSelectedItem();
    if(treeCtrl.ItemHasChildren(hItem))return;                    //若非功能结点,则返回

    CString strChild=treeCtrl.GetItemText(hItem);
                                                //获得叶结点字符串,相当于 ini 中的 key
    HTREEITEM hParent=treeCtrl.GetParentItem(hItem);
    CString strParent=treeCtrl.GetItemText(hParent);
                                                //获得根结点字符串,相当于 ini 中的 section
    CMyProjDoc * pDoc= (CMyProjDoc *)GetDocument();
    pDoc->FindProc(strParent,strChild);         //根据 section,key 进行查询
    * pResult=0;
}
```

(8) 文档类 CMyProjDoc 的编写。

```cpp
//MyProjDoc.h 头文件
class CMyProjDoc : public CDocument
{
protected:                                      //create from serialization only
    CMyProjDoc();
    DECLARE_DYNCREATE(CMyProjDoc)
public:
    MySectionCollect m_collect;                 //定义 ini 文件映射对象
    CContentView * m_pContentView;              //树型视图指针,用于反映查询内容
    CMyProjView * m_pDataView;                  //列表视图指针,用于反映查询结果
public:
    BOOL FillContentView();                     //填充树型视图,即查询内容
    BOOL AddItem(int nRow,int nCol,CString strText);   //列表视图添加项用到的函数
    void FillDataList(string strSQL);           //填充查询结果至列表控件
    BOOL FindProc(CString strParent,CString strChild); //完成查询主函数
//Overrides
    //{{AFX_VIRTUAL(CMyProjDoc)
public:
    virtual BOOL OnNewDocument();
    //}}AFX_VIRTUAL
public:
```

```cpp
    virtual ~CMyProjDoc();
    DECLARE_MESSAGE_MAP()
};

//MyProjDoc.cpp 源文件
IMPLEMENT_DYNCREATE(CMyProjDoc,CDocument)
BEGIN_MESSAGE_MAP(CMyProjDoc,CDocument)
END_MESSAGE_MAP()

CMyProjDoc::CMyProjDoc()
{}
CMyProjDoc::~CMyProjDoc()
{}
BOOL CMyProjDoc::FillContentView()                        //填充树型视图,即查询内容
{                                      //假想就两层,第一层是查询分类描述,第二层是查询项描述
    CTreeCtrl & treeCtrl=m_pContentView->GetTreeCtrl();

    MySection * pSection=m_collect.GetSection("mycontent");
    map<string,string>::iterator it=pSection->mapKey.begin();
    while(it!=pSection->mapKey.end())
    {
        HTREEITEM hItem=treeCtrl.InsertItem((*it).first.c_str());
        string strItems=(*it).second;

        int first=0,last=0;
        string s;
        if(strItems.find(",")!=string::npos)
        {
            while((last=strItems.find_first_of(",",first))!=string::npos)
            {
                s=strItems.substr(first,last-first);
                treeCtrl.InsertItem(s.c_str(),hItem);
                first=last+1;
            }
            s=strItems.substr(first,strItems.length()-first);
            treeCtrl.InsertItem(s.c_str(),hItem);
        }
        it++;
    }
    return TRUE;
}
```

```cpp
BOOL CMyProjDoc::AddItem(int nRow,int nCol,CString strText)   //列表视图添加项用到的函数
{
    CListCtrl & listCtrl=m_pDataView->GetListCtrl();

    USES_CONVERSION;
    LVITEM lvItem;
    lvItem.mask     =LVIF_TEXT;
    lvItem.iItem    =nRow;
    lvItem.iSubItem =nCol;
    lvItem.pszText  = (LPTSTR)(LPCTSTR)strText ;

    if(nCol==0)
        return listCtrl.InsertItem(&lvItem);
    else
        return listCtrl.SetItem(&lvItem);
    return TRUE;
}
void CMyProjDoc::FillDataList(string strSQL)              //填充查询结果至列表控件
{
    string connect=m_collect.Find("mydatabase","connect");
    string user=m_collect.Find("mydatabase","user");
    string pwd=m_collect.Find("mydatabase","pwd");

    CAdoFunc myado;
    myado.Connect(connect.c_str(),"","");
    myado.Execute(strSQL.c_str());
    if(myado.IsEmpty())
    {
        myado.Close();
        return;
    }

    CListCtrl & listCtrl=m_pDataView->GetListCtrl();
    CMyTable myTable(&myado,"");
    myTable.ExecuteQuery(strSQL.c_str());                 //执行查询
    //删除表头及数据
    CHeaderCtrl * phead=listCtrl.GetHeaderCtrl();
    int nCols=phead->GetItemCount();
    if(nCols>=1)
        while(listCtrl.DeleteColumn(0));
    listCtrl.DeleteAllItems();
```

```cpp
//填充表头
LVCOLUMN lc;
lc.mask=LVCF_TEXT|LVCF_WIDTH;
lc.cchTextMax=100;
vector<CMyField>& vecField=myTable.m_vecField;
for(int i=0; i<vecField.size(); i++)
{
    lc.cx=listCtrl.GetStringWidth(vecField[i].strFieldName)+15;
    lc.pszText=(LPTSTR)(LPCTSTR)vecField[i].strFieldName;
    listCtrl.InsertColumn(i,&lc);
}

//填充表中数据
deque<deque<CString>>& d=myTable.m_vecData;
for(i=0; i<d.size(); i++)
{
    for(int j=0; j<d[0].size(); j++)
    {
        AddItem(i,j,d[i][j]);
    }
}
myado.Close();
}
BOOL CMyProjDoc::FindProc(CString strParent,CString strChild)    //完成查询主函数
{
    string parent=strParent,child=strChild;
    string s=m_collect.Find(parent,child);
    CFindPara obj(parent,child,s);
    CFindDlg dlg(&obj);
    if(dlg.DoModal()==IDOK)
    {
        s=obj.MergeSQL(dlg.m_vecResult);
        FillDataList(s);
    }
    return TRUE;
}
BOOL CMyProjDoc::OnNewDocument()
{
    if(!CDocument::OnNewDocument())
        return FALSE;
    string path="d:\\MyConfig.ini";
    ReadIni ri(path,m_collect);                        //读ini文件类初始化
```

```
        ri.Process();                              //读ini文件过程
        return TRUE;
}
```

可以看出,文档类是一个与数据相关的中心类,数据指的是配置文件的内容,因此有一个成员变量 MySectionCollect m_collect,用以把配置文件内容以一定的数据结构保存到内存中。由于要把 m_collect 的一部分内容映射成树型视图,一部分内容映射成列表视图,因此有两个指针视图成员变量 m_pContentView 及 m_pDataView,分别对应树型视图和列表视图。

当主界面图 11.12(a)显示前,程序主要流程如下所示。

(1)调用文档类 CMyProjDoc 类中的 OnNewDocument 函数,在该函数内完成读取 MyConfig.ini 文件功能,保存在 m_collect 成员变量中。

(2)初始化列表视图类 CMyProjView,并在该类的 OnInitialUpdate 函数中给文档类列表视图指针成员变量 m_pDataView 赋值。

(3)初始化树型视图类 CContentView,并在该类的 OnInitialUpdate 函数中给文档类树型视图指针成员变量 m_pContentView 赋值,并且调用文档类中的 FillContentView 函数填充树型视图,这时就出现了图 11.12(a)所示界面。

当主界面出现后,有效程序流程如下所示。

(1)双击树型视图叶结点(查询结点),响应 CContentView 中 OnDblclk 函数,在该函数内调用文档类 FindProc 函数。

(2)在 FindProc 函数内,先生成查询参数 CFindPara 对象,然后依据该对象动态生成 CFindDlg 对象,填充查询信息,最后形成完整的 SQL 查询语句,调用文档类 FillDataList 函数填充列表视图的查询结果。

也许有同学说,配置文件采用 XML 更好,这是毫无疑问的。这与本示例论述的重点并不矛盾,本示例旨在说明 STL 是非常适于封装配置文件内容的,无论是 ini、XML 还是其他格式。只是由于 ini 格式比较简单,精通了 STL 在 ini 格式文件上的用法,那么扩展至 XML 格式也就容易了。

11.3.5 STL 与动态链接库

【例 11.18】 STL 在动态链接库中的应用。

以上的所有示例都是在一个应用程序中完成的,而当前编程模式往往是应用程序+动态链接库的形式,那么能否说把单应用程序模式经过简单的工作就能转化成应用程序+动态链接库的形式呢?事情不像同学们想像的那样简单。关键是动态链接库接口怎样定义?接口函数参数能是任意类型吗?内存分配原则如在应用程序分配,在动态链接库中删除,这样好不好等。当前比较流行的是 ATL 动态链接库,ATL 的主要目的是创建小的,基于 COM 的软件模块,然后再把这些模块组装成大的应用程序。COM 是程序开发的方向,ATL 则是开发者手中最为重要的工具。ATL 动态链接库一般要求接口参数是基本数据类型或基本数据类型组成的结构体,如表 11.8 所示。

表 11.8 动态链接库接口函数一般参数类型

基本类型	说 明	基本类型	说 明
boolean	8 位数据项	int	32 位符号整型
byte	8 位数据项	long	32 位符号整型
char	8 位无符号数据项	short	16 位符号整型
double	64 位浮点数据	small	8 位符号整型
float	32 位浮点数据	void *	32 位环境句柄指针类型
handle_t	原始句柄类型	wchar_t	16 位无符号数据项
hypersigned	64 位符号整型	BSTR	带长度前缀的字符串

因此,动态链接库中一般不传递对象参数,否则尽管编译通过,但执行时容易引起 Access violation 等异常。例如,若单应用程序由主程序 A+类 B 组成,那么在 A 中很容易就得到 B 对象的引用;但如果 B 在动态链接库中,那么 A 很难得到 B 对象的引用。

例如,例 11.3 中功能是解析 ini 格式文件,在 main 函数中可以很方便地得到 MySectionCollect 对象的信息,从而获得 ini 配置文件中的信息。但现在如果把读 ini 文件功能即把 MySection、MySectionCollect 和 ReadIni 三个类封装成 ATL 动态链接库,应该如何编制各接口函数才能方便调用者获得配置文件信息呢?下面就介绍这一编程经过。

(1) 产生新的 ATL COM AppWizard 工程,工程名选为 MySub。

(2) 添加例 11.3 中 MySection、MySectionCollect 和 ReadIni 三个类。

(3) 选择 Insert 菜单中的 New ATL Object 命令,选择 Simple Object,输入接口类名 MyConfig,单击"确定"按钮即可,如图 11.15 所示。

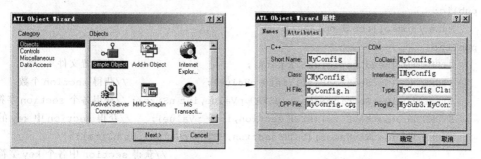

图 11.15 操作界面

(4) 在 class view 窗口中选择接口 IMyConfig,右击选择 Add Method 命令,依次输入 6 个接口函数。in 和 out 关键字指明了参数传递的方向,in 表示输入,out 表示输出,一般是指针类型。参考操作界面如图 11.16 所示。

产生的头文件及源文件如下所示。

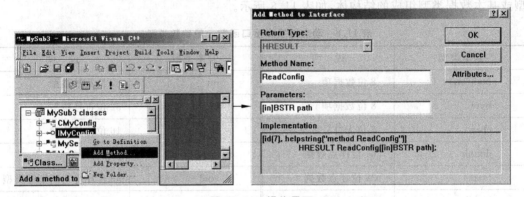

图 11.16　操作界面

```
//MyConfig.h 头文件
class ATL_NO_VTABLE CMyConfig : public CComObjectRootEx<CComSingleThreadModel>,
    public CComCoClass<CMyConfig,&CLSID_MyConfig>,
    public IDispatchImpl<IMyConfig,&IID_IMyConfig,&LIBID_MYSUB3Lib>
{
public:
    CMyConfig()
    {}
DECLARE_REGISTRY_RESOURCEID(IDR_MYCONFIG)
DECLARE_PROTECT_FINAL_CONSTRUCT()
BEGIN_COM_MAP(CMyConfig)
    COM_INTERFACE_ENTRY(IMyConfig)
    COM_INTERFACE_ENTRY(IDispatch)
END_COM_MAP()
public:
    MySectionCollect m_collect;
public:
    STDMETHOD(ReadConfig)(BSTR path);                               //读配置文件
    STDMETHOD(GetSectionSize)(int * pValue);                        //获得 section 个数
    STDMETHOD(GetSectionNames)(BSTR * pValue,int nSize); //获得各个 section 字符串
    STDMETHOD(GetKeySize)(BSTR section,int * pValue);    //获得 section 中 key 的个数
    STDMETHOD(GetKeyNames)(BSTR section,BSTR * pValue,int nSize);
                                                //获得 section 中各个 key 字符串值
    STDMETHOD(GetValue)(BSTR section,BSTR key,BSTR * pValue);
                                                //获得 section+key 对应的 value
};

//MyConfig.cpp 源文件
STDMETHODIMP CMyConfig::ReadConfig(BSTR path)
{
```

```
    USES_CONVERSION;
    ReadIni ri(OLE2T(path),m_collect);        //读ini文件类初始化
    ri.Process();                              //读ini文件过程
    return S_OK;
}
STDMETHODIMP CMyConfig::GetSectionSize(int * pValue)
{
    map<string,MySection> * pMap=m_collect.GetCollect();
    * pValue=pMap->size();
    return S_OK;
}
STDMETHODIMP CMyConfig::GetSectionNames(BSTR * pValue,int nSize)
{
    USES_CONVERSION;
    int pos=0;
    map<string,MySection> * pMap=m_collect.GetCollect();
    map<string,MySection>::iterator it=pMap->begin();
    while(it!=pMap->end()&& pos<nSize)
    {
        pValue[pos]=T2BSTR((* it).first.c_str());
        pos++;
        it++;
    }
    return S_OK;
}
STDMETHODIMP CMyConfig::GetKeySize(BSTR section,int * pValue)
{
    USES_CONVERSION;
    MySection * pSection=m_collect.GetSection(OLE2T(section));
    * pValue=pSection->mapKey.size();
    return S_OK;
}
STDMETHODIMP CMyConfig::GetKeyNames(BSTR section,BSTR * pValue,int nSize)
{
    USES_CONVERSION;
    MySection * pSection=m_collect.GetSection(OLE2T(section));
    map<string,string>& m=pSection->mapKey;
    int pos=0;
    map<string,string>::iterator it=m.begin();
    while(it!=m.end()&& pos<nSize)
    {
        pValue[pos]=T2BSTR((* it).first.c_str());
        pos++;
```

```
        it++;
    }
    return S_OK;
}
STDMETHODIMP CMyConfig::GetValue(BSTR section,BSTR key,BSTR * pValue)
{
    USES_CONVERSION;
    string sec=OLE2T(section);
    string k   =OLE2T(key);
    string value =m_collect.Find(sec,k);
    * pValue=T2BSTR(value.c_str());
    return S_OK;
}
```

可以看出，ReadConfig 函数是完成读配置文件功能的，操作的仍是 STL 基础类 MySection、MySectionCollect、ReadIni。那么头文件中定义的 5 个其他接口函数是如何分析定义的呢？从语义角度思考，要想方便获知配置文件的内容，就应该知道配置文件中有多少个 section，每个 section 的标识字符串是什么，每个 section 中有多少个 key，每个 key 的标识字符串是什么，每个 section+key 对应的 value 是什么。转化成计算机描述方式，就是头文件中定义的其他 5 个函数。由于动态链接库接口函数参数类型的限制性，这在一定程度上决定了接口功能都很细化，可能有的同学初学的时候感到非常啰唆，但却非常方便组合，可以得到所需的任何内容，这也正是动态链接库的一大特点。它反过来也指导我们编制普通类时，也尽量做到细化，加深对类的理解。用通俗一点的话来讲，类是"分类"的意思，不要在一个"分类"中包含多个"分类"的内容。

示例动态链接库编译链接完成后，其中自动生成了两个有用的文件 MySub_i.c 及 MySub.h，主要是客户端调用需要这两个文件，其内容如下所示。

```
//MySub_i.c 主要内容
typedef struct _IID
{
    unsigned long x;
    unsigned short s1;
    unsigned short s2;
    unsigned char c[8];
}IID;

const IID IID_IMyConfig = {0x02A8F1F8, 0x140A, 0x4840, {0xAE, 0xA1, 0x19, 0xE9, 0xC7, 0x24, 0xEC, 0xFD}};
constIID LIBID_MYSUBLib=
{0x108F731B,0xC768,0x4E4F,{0xA2,0x18,0xA0,0x71,0xAF,0x7A,0xCC,0x96}};
const CLSID CLSID_MyConfig=
```

{0xFAC2AFA7,0x9314,0x4B6D,{0xBB,0x83,0x72,0x72,0x60,0xC3,0x92,0x59}}};

```
//MySub.h 主要内容
IMyConfig:public IDispatch
{
public:
    virtual HRESULT STDMETHODCALLTYPE ReadConfig(BSTR path)=0;
    virtual HRESULT STDMETHODCALLTYPE GetSectionSize(int __RPC_FAR * pValue)=0;
    virtual HRESULT STDMETHODCALLTYPE GetSectionNames(BSTR __RPC_FAR * pValue,int nSize)=0;
    virtual HRESULT STDMETHODCALLTYPE GetKeySize (BSTR section, int __RPC _FAR * pValue)=0;
    virtual HRESULT STDMETHODCALLTYPE GetKeyNames (BSTR section,BSTR __RPC_FAR * pValue,int nSize)=0;
    virtual HRESULT STDMETHODCALLTYPE GetValue (BSTR section,BSTR key,BSTR __RPC_FAR * pValue)=0;
};
```

MySub_i.h 定义了接口类的 uuid 标识符,用于向注册表注册；MySub.h 定义了 6 个接口函数的虚函数类。也就是说,应用程序根据 MySub_i.c 中的 CLSID_MyConfig、IID_IMyConfig 通过注册表即可获得动态链接库的 IMyConfig 接口,通过该接口就可以操作其所具有的 6 个函数功能了。

下面看一个测试程序执行界面,如图 11.17 所示。

图 11.17 应用程序调用动态链接库界面示例

初始化时填充 section 列表框,表明有多少 section。当双击某 section 时,填充该 section 中包含的 key 内容至 key 列表框,当双击 key 列表框中某项时,相当于 section、key 都确定了,则在下方编辑框中显示出 value 值。

编程步骤如下所示。

(1) 建立对话框应用程序,工程名为 MyApp。

(2) 复制粘贴 MySub 工程中的 MySub_i.c、MySub.h 文件,并添加至 MyApp 工程中。

(3) 完善 CMyAppDlg,头文件及源文件代码如下所示。

```
//MyAppDlg.h 头文件主要内容
class CMyAppDlg : public CDialog
{
    DECLARE_DYNAMIC(CMyAppDlg);
//Construction
public:
    CMyAppDlg(CWnd* pParent=NULL);          //standard constructor
    virtual~CMyAppDlg();
//Dialog Data
    //{{AFX_DATA(CMyAppDlg)
    enum{IDD=IDD_MYAPP_DIALOG};
    CListBox    m_sectlist;                 //section 列表框
    CListBox    m_keylist;                  //key 列表框
    CString     m_value;                    //value 编辑框
    //}}AFX_DATA
    IMyConfig* m_pConfig;                   //动态链接库接口
    //ClassWizard generated virtual function overrides
    //{{AFX_VIRTUAL(CMyAppDlg)
    protected:
    virtual void DoDataExchange(CDataExchange* pDX);   //DDX/DDV support
    //}}AFX_VIRTUAL
protected:
    //Generated message map functions
    //{{AFX_MSG(CMyAppDlg)
    virtual BOOL OnInitDialog();            //填充 section 列表框
    afx_msg void OnFillKey();               //填充 key 列表框
    afx_msg void OnFillValue();             //填充 value 编辑框
    afx_msg void OnDestroy();
    //}}AFX_MSG
    DECLARE_MESSAGE_MAP()
};


//MyAppDlg.cpp 源文件
IMPLEMENT_DYNAMIC(CMyApp3Dlg,CDialog);
CMyApp3Dlg::CMyApp3Dlg(CWnd* pParent/*=NULL*/): CDialog(CMyApp3Dlg::IDD,pParent)
{
    m_value=_T("");
    m_pConfig=NULL;
}
CMyApp3Dlg::~CMyApp3Dlg()
{}
void CMyApp3Dlg::DoDataExchange(CDataExchange* pDX)
{
    CDialog::DoDataExchange(pDX);
```

```cpp
    //{{AFX_DATA_MAP(CMyApp3Dlg)
    DDX_Control(pDX,IDC_LIST2,m_keylist);
    DDX_Control(pDX,IDC_LIST1,m_sectlist);
    DDX_Text(pDX,IDC_EDIT1,m_value);
    //}}AFX_DATA_MAP
}

BEGIN_MESSAGE_MAP(CMyApp3Dlg,CDialog)
    //{{AFX_MSG_MAP(CMyApp3Dlg)
    ON_LBN_DBLCLK(IDC_LIST1,OnFillKey)
    ON_LBN_DBLCLK(IDC_LIST2,OnFillValue)
    ON_WM_DESTROY()
    //}}AFX_MSG_MAP
END_MESSAGE_MAP()

BOOL CMyApp3Dlg::OnInitDialog()
{
    CDialog::OnInitDialog();

    USES_CONVERSION;
    CoInitialize(NULL);                              //初始化COM库,创建IMyConfig接口实例
     CoCreateInstance(CLSID_MyConfig,NULL,CLSCTX_INPROC_SERVER,IID_IMyConfig,
(LPVOID*)&m_pConfig);
    CComBSTR bstrPath2("d:\\MyConfig.ini");
    m_pConfig->ReadConfig(bstrPath2);                //读配置文件

    int nSize=0;                                     //填充section列表框
    m_pConfig->GetSectionSize(&nSize);
    CComBSTR *pBstr=new CComBSTR[nSize];
    m_pConfig->GetSectionNames((BSTR*)pBstr,nSize);
    for(int i=0; i<nSize; i++)
    {
        m_sectlist.AddString(OLE2T(pBstr[i]));
    }
    delete []pBstr;
    return TRUE;
}
void CMyApp3Dlg::OnFillKey()                         //填充key列表框
{
    m_keylist.ResetContent();                        //清空key列表框
    int index=m_sectlist.GetCurSel();                //获得section串
    CString str;
    m_sectlist.GetText(index,str);
```

```cpp
        USES_CONVERSION;

        int nKeySize=0;                                    //根据 section 串确定有多少 key
        m_pConfig->GetKeySize(T2BSTR(str),&nKeySize);
        CComBSTR *pSubstr=new CComBSTR[nKeySize];          //利用数组获得各 key 串内容
        m_pConfig->GetKeyNames(T2BSTR(str),(BSTR * )pSubstr,nKeySize);
        for(int i=0; i<nKeySize; i++)
        {
            m_keylist.AddString(OLE2T(pSubstr[i]));
        }
        delete []pSubstr;
    }
    void CMyApp3Dlg::OnFillValue()                         //填充 value 编辑框
    {
        CString sect;
        int index=m_sectlist.GetCurSel();
        m_sectlist.GetText(index,sect);                    //获得 section

        CString key;
        index=m_keylist.GetCurSel();
        m_keylist.GetText(index,key);                      //获得 key

        USES_CONVERSION;
        CComBSTR value;                                    //获得 value 并填充编辑框
        m_pConfig->GetValue(T2BSTR(sect),T2BSTR(key),&value);
        m_value=OLE2T(value);
        UpdateData(FALSE);
    }
    void CMyApp3Dlg::OnDestroy()
    {
        CDialog::OnDestroy();
        m_pConfig->Release();                              //释放动态链接库资源
        m_pConfig=NULL;
        CoUninitialize();                                  //释放 COM 库资源
    }
```

执行 ATL 动态链接库的一般步骤是：

（1）初始化 COM 库，示例中是在 OnInitDialog 函数中完成的，即调用 CoInitialize(NULL)；

（2）获得 COM 接口，本程序即 IMyConfig 接口 m_pConfig，它是通过 CoCreateInstance 函数完成的；

（3）通过 m_pConfig 操纵所需各接口函数；

（4）在程序适当位置删除 COM 资源，删除 COM 库资源，本程序是在 OnDestroy 函数中完成的。

参 考 文 献

[1] Bruce Eckel,Chuck Allison. C++编程思想：第2卷 实用编程技术. 刁成嘉译. 北京：机械工业出版社,2006.
[2] 侯捷. STL源码剖析. 武汉：华中科技大学出版社,2002.
[3] 叶至军. C++ STL开发技术导引. 北京：人民邮电出版社,2007.
[4] 彭木根,王淑凌. C++ STL程序员开发指南. 北京：中国铁道出版社,2003.
[5] Plauger P J,Alexander A Stepanov,Meng Lee. C++ STL中文版. 王昕译. 北京：中国电力出版社,2002.
[6] Herb Sutter. Exception C++中文版. 桌小涛译. 北京：中国电力出版社,2003.
[7] Herb Sutter. MoreException C++中文版. 於竞春译. 武汉：华中科技大学出版社,2002.
[8] M. H. Alsuwaiyel. 算法设计技巧与分析. 吴伟昶,方世昌等译. 北京：电子工业出版社,2004.
[9] Tom Armstrong,Ron Patton. ATL开发指南. 董梁,丁杰,李长业译. 北京：电子工业出版社,2000.
[10] 殷人昆,陶永雷,谢若阳. 数据结构(用面向对象方发与C++描述). 北京：清华大学出版社,1999.
[11] 严蔚敏,吴伟民. 数据结构(C语言版). 北京：清华大学出版社,2007.
[12] William J. Collins. 数据结构与STL. 周翔译. 北京：机械工业出版社,2004.
[13] 王晓东. 数据结构(STL框架). 北京：清华大学出版社,2009.
[14] Scott Meyers. Effective STL中文版. 潘爱民译. 北京：清华大学出版社,2006.
[15] 刘德山,金百东. Java程序设计[M]. 北京：科学出版社,2012.
[16] 刘德山,金百东. Java设计模式深入研究[M]. 北京：人民邮电出版社,2014.
[17] 刘德山,杨彬彬. 基于Hibernate数据持久层架构设计与应用[J]. 微型机与应用,2011(8)：12-16.
[18] 郭瑾,高伟,刘德山. 基于Ajax的网络答疑系统的研究与应用[J]. 微型电脑应用,2010(5) 39-42.
[19] 金百东,刘德山,. Java程序设计学习指导与习题解答.[M]. 北京：科学出版社,2012.
[20] 金百东,李文举. 利用C++ ATL技术实现反射机制[J]. 微型机与应用,2013(2)：4-6.
[21] 金百东. MIS系统通用报表的设计与实现[J]. 计算机与数字工程,2005(6)：116-119.
[22] 金百东. ERP复合查询显示组件设计与实现[J]. 自动化技术与应用,2008(12)：59-62.
[23] 金百东. 基于MetaFile图元文件的数据集成显示与打印[J]. 计算机应用与软件,2007(6)：93-94.